Yale Egyptological Studies 10

# Feasts and Fights:
# Essays on Time in Ancient Egypt

Yale Egyptological Studies

Yale Egyptological Studies 10

# Feasts and Fights

## Essays on Time in Ancient Egypt

Anthony Spalinger

Yale Egyptological Institute New Haven, CT

HAYIM TADMOR

*(18/11/1923: Harbin; 11/12/2005: Jerusalem)*

Yale Egyptological Studies 10

ISBN 978-0-9740025-8-3



*Library of Congress Cataloging-in-Publication Data*

Names: Spalinger, Anthony John, author.
Title: Feasts and Fights : Essays on Time in Ancient Egypt / Anthony Spalinger.
Other titles: Yale Egyptological studies ; 10.
Description: New Haven, CT : Yale Egyptological Institute, 2017. | Series: Yale Egyptological studies ; 10 | Includes bibliographical references and index.
Identifiers: LCCN 2017036456 | ISBN 9780974002583 (pbk. : alk. paper)
Subjects: LCSH: Calendar, Egyptian. | Calendar--Egypt--History. | Egypt--History--To 332 B.C. | Chronology, Historical. | Kadesh, Battle of, Kadesh, ca. 1300 B.C. | Egypt--History--Nineteenth dynasty, ca. 1320-1200 B.C.
Classification: LCC CE29 .S69 2017 | DDC 529/.32932--dc23
LC record available at https://lccn.loc.gov/2017036456



Typeset in Minion Pro and Warnock Pro.

Printed in the United States of America on acid-free paper.

# Contents

# List of Illustrations

# Editor's Preface

During a seminar held in the Department of Near Eastern Languages and Civilizations at Yale University during January of 2012, Prof. Tony Spalinger presented the contents of this volume, in many ways a summation and expansion of much of his previous work on the calendars of ancient Egypt. I am pleased that we can now see these lectures published in the *Yale Egyptological Studies*, a volume that furthers the work of Prof. Spalinger on calendrical and military matters.

Proceeding from a presentation of the calendrical aspects of Egyptian festivals—already an impressive and imminently helpful summation and clarification—Prof. Spalinger proceeds to engage with an important earlier work, the late Prof. Richard Parker's *The Calendars of Ancient Egypt*, cogently explaining both the important contributions and remaining flaws in that small magnum opus. For many, the most important aspect of Prof. Spalinger's volume will be his reconstruction of the development of a new Civil Calendar, and concomitant transformation of Egyptian dating systems during the transition between the Early Dynastic Period and the high Old Kingdom. The final chapters both continue the volume's focus on temporal matters, and combine that with the author's long-standing interest and expertise in military historical topics.

John Coleman Darnell
New Haven, CT

# Preface

The following chapters were presented in the Department of Near Eastern Languages and Civilizations at Yale University from January 23 to 27, 2012. I am in great debt to the two Egyptologists at my former graduate school, Professors John Darnell and Colleen Darnell, who invited me to deliver lectures on these subjects, fruits of a long period of research. I had thought it best to commence with a detailed study of the interrelations between Egyptian festivals and their temporal settings, since others, especially Ronald Wells, had urged me to draw up a "final" report of past research on this topic, and it seemed to me that a summation, or a capstone, of earlier endeavors would prove worthwhile.

It is odd, perhaps uncanny, to be placed in an apparent, though not dramatic, situation of anxiety when one is asked to state "what do you *really* believe?" Hence, the first chapter in this volume is the longest, even if it opens with few presuppositions. Nevertheless, at the beginning I wish to make my perspective clear. I do not believe that ancient Egyptian calendrics and chronology are recondite matters. It is that we moderns have made it so. Intricate charts, often lacking any statistical awareness of error parameters, are frequently placed before the unsuspecting reader, whereas the foundations of the subject remain inconspicuous. Hence, Chapter 1 has provided the exhausted or blasé student with the basic information and a set of appropriate data. This way, the material will be more accessible and less off-putting than quite a number of Egyptological publications concerned with time, either abstractly or otherwise.[1]

The second and third chapters move to the heart of the matter. Yet I have not neglected to present the difficulties inherent in one classic of the field, Richard A. Parker's *The Calendars of Ancient Egypt*.[2] I considered it a desirable necessity to use his erudite 1950 publication as a basis, and not merely as a skeleton, for subsequent work. Therefore, I have guided the reader to see how well-developed and advanced his hypotheses were, especially in comparison to those of Heinrich Brugsch and

1 See my review of Jan Assmann, *Steinzeit und Sternzeit: Altägyptische Zeitkonzepte* (Munich, 2011), *WZKM* 102 (2012): 303–10.

2 Richard A. Parker, *The Calendars of Ancient Egypt* (Chicago, 1950).

Ludwig Borchardt. But my purpose was also to explain the difficulties that Parker faced with his salutary attempt to clarify eponymous feasts of ancient Egypt. In many ways, that portion of this work is partly based on a recent drawing together of disparate strands of scholarship, including the pertinent remarks of Rolf Krauss,[3] in an effort to determine the operation of the religious celebrations that gave their names to the primitive lunar months. Hence, Chapter 2 is partly dependent upon my recent study, "The Beginning of the Civil Calendar," which was written at Prague in 2009 and presented during the symposium *Abusir and Saqqara in 2010*.[4]

In that study, I set out to extend my arguments for the non-existence of any so-called "second lunar calendar."[5] I had earlier removed this hypothesis of Parker without, however, replacing it with any explanation why the eponymous feasts "bobbed" back and forth and why their names were later applied to the civil months by a simple, but often ignored rule of Y = X-1, where X is the old name of the lunar month and Y is its transference into the subsequent Civil Calendar. If a straightforward explanation could be found, then there was no need to encumber Egyptology with hyper-baroque reconstructions of nomenclature, dating, and societal transformation. That I ultimately focused upon the reign of Djoser would probably come as no surprise to Wolfgang Helck, who repeatedly stressed the phenomenal transformation of the archaic society of Egypt into a fully mature early state during that king's reign. But it was Oleg Berlev who led me in this quest.[6] If one follows Parker's careful argumentation as well, this era fits into his invention of the now discredited "second lunar calendar," as chronologically argued in 1949–50. Note that I do not necessarily follow Berlev in his interpretation of a statue of Djoser found at Saqqara, and the present chronological reconstruction of Egyptologists, myself included, might regard the date as *sub judice*.

3 Rolf Krauss, *Sothis- und Monddaten: Studien zur astronomischen und technischen Chronologie Altägyptens* (Hildesheim, 1985). I find that Danijela Stefanovic, "The Middle Kingdom Chronology: Present Knowledge and Confusions," *Journal of the Serbian Archaeological Society* 17 (2001) gives a very honest and judicious analysis. Wells would agree with her. Neither he nor I, however, would concur with the study of Boris Banjevic, "Lunar Dates and Chronology of the Late Middle Kingdom of Ancient Egypt," on pp. 105–14 in the same journal.

4 The official title was "International Workshop: Abusir and Saqqara in the Year 2010: Prague, May 31–June 4 2010." The final publication is "The Beginning of the Civil Calendar," in Miroslav Bárta, Filip Coppens, and Jaromír Krejčí (eds.), *Abusir and Saqqara in the Year 2010/2* (Prague, 2012), 723–35.

5 Parker presented this daring hypothesis in his *Calendars*. The presumed evidence was reinterpreted by me in "Month Representations," *CdÉ* 70 (1995): 110–22.

6 Oleg Berlev, "Precisely Two Sothic Periods Between Year 18 of King *Sénw*/Tosorthros and Year 2 of Pharaoh Antoninus Pius," in O. I. Pavlova, ed., *Ancient Egypt: Language, Culture, Consciousness* (Moscow, 1999), 42–62.

The next chapter presents the results of the inquiry. With the basis of the new Civil Calendar explicated, everything falls into place. No complicated series of causal requirements needs be posited. No arithmetic calculations have to be scanned, digested, and then memorized. As befits the nature of Egyptian civilization between ca 3100 and 2700 BC, the transformation of its calendars was relatively transparent, to them as it should be to us. It is for this reason that I have devoted more time to the global results of our research than to any specific religious celebration. The conclusions to the entire project of calendrical reorientation should enable one to see how futile it is to search for any complicated ancient system of temporal reckoning. The Egyptian calendrical world was not an Arcanum. Attempts otherwise seem to be like the proverbial month of March, entering ferociously as a lion and leaving, very subdued, as a lamb.

The final two chapters change our direction and perhaps for some, we are advancing windward. I am sure that it is not surprising to many that I have chosen two military subjects to round off this small volume. Yes, once again Kadesh is discussed, but now in the light of a tightly-knit temporal analysis of the logistics of battle.[7] Even here, I can frankly state, time keeping still applies. This study, as well as the final one that deals with the iconography of Seti I's Karnak reliefs, also continues earlier research into New Kingdom warfare and especially its pictorial representations. The last chapter, in fact, is linked with a recent publication concerning alterity and the images or icons that the Egyptian felt necessary to use in order to present their feelings — and strong they were! — about the Other.[8]

Even in the Seti chapter there is the backgrounding of a temporal unity, one that runs through this entire volume. Hence, I do not feel it necessary to defend my combination of calendars and warfare. Both are interconnected, and to unravel the complexities of time at the Battle of Kadesh is as demanding as to redraw the Egyptian system of eponymous feasts and month names.

I had great pleasure in presenting this work at Yale University and hope that I have made myself understood to the most demanding critic. To offer these topics to a mature, enlightened, and receptive audience was one of the personal highlights of the Yale experience that cannot be repeated in written format. Yet, if granted more years, I will endeavor to continue with this direction of research.

The dedication is to one remarkable visiting professor in the then Department of Near Eastern Languages and Literatures with whom I had the great pleasure to work during my Yale years.

Anthony Spalinger
31 December 2016

---

7 Brett H. Heagren, *The Art of War in Pharaonic Egypt: An Analysis of the Tactical, Logistic, and Operational Capabilities of the Egyptian Army (Dynasties XVII–XX)* (Auckland, PhD Thesis, 2010).

8 Anthony Spalinger, *Icons of Power: A Strategy of Reinterpretation* (Prague, 2011).

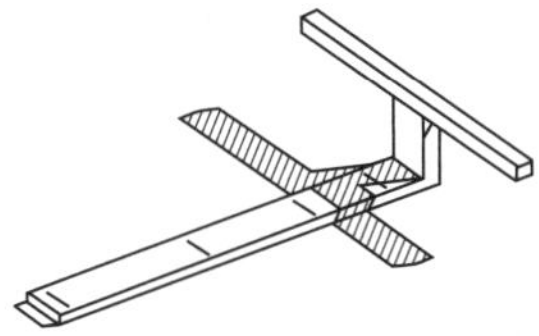

# 1
# Calendrics: An Introduction

## A. BACKGROUND

The Egyptian calendar has often been called the sanest method of chronological measurement that has ever been devised.[1] The 365-day year is as arithmetically simple as it is close in duration to the actual longitudinal path of the sun in one (solar) year. That the Egyptians selected a non-lunar basis for their annual measurement of time distinguishes them equally from most other contemporary civilizations, such as Mesopotamia, as well as from medieval cultures. When we read Egyptian texts that include dates, the first thing that we notice is how ancient and extensive this set-up was.

The calendrical system employed in historical times was an annual year comprised of 12 months, each having 30 days, and 5 extra days at the end of the year.[2] The latter are given the Greek term of epagomenal days, although the native expression was "above the year," or later, in Coptic, as a "short month."[3] The radical simplicity seems to have come to pass in early historical time. I would place the commencement of what

1 Otto Neugebauer, *The Exact Sciences in Antiquity*[2] (New York, 1969), p. 81.

2 For an overview, see Rolf Krauss, *Sothis- und Monddaten: Studien zur astronomischen und technischen Chronologie Altägyptens* (Hildesheim, 1985), pp. 11–14. Marshall Clagett, *Ancient Egyptian Science: Calendars, Clocks, and Astronomy* (Philadelphia, 1995), can be as frustrating as it is useful. Part of the difficulty in using this work is that the author had not investigated the secondary source material since ca. 1989 and depended too much upon the "standard" texts and scenes from pharaonic Egypt. Quite an amount of primary data was overlooked. Nevertheless, he saw the inherent difficulties in Parker's analysis — see note 6, below.

3 Anthony Spalinger, "Some Remarks on the Epagomenal Days in Ancient Egypt," *JNES* 54 (1994): 33–47; with Graeme Harrison, "The Position of the Day *grḥ nẖnw m sš.f* — 'The night the Child is in his Nest' within the Epagomenal Period," *GM* 143 (1994): 77–79. Add Didier Devauchelle, "Écrire le nom des jours épagomènes et du premier jour de l'an [O. Dém Delm 4–1]," *SEP* 2 (2005): 75–81. Danielle Bonneau, "Les fêtes

is called the civil year of 365 days to the Third Dynasty, and under pharaoh Djoser. (See our later discussion in Chapter 2.) It is apparent in the historical records of the Old Kingdom that this year was in operation. Although labeled by most scholars as the Civil Calendar, some have returned to the less useful Latin phrase of *annus vagus*.[4] Various legal documents of the late Old Kingdom, as well as numerous royal inscriptions (in the Fourth to the Sixth Dynasties) all attest to the effective working of the Civil Calendar.

The origins of this calendar are controversial. It was earlier argued that the 365-day length was derived from the annual reappearance of the star Sothis (Latin Sirius) after a period of invisibility of 70 days.[5] Because Sothis played such an important

d'Amesysia," *CdÉ* 49 (1974): 366–79 covers, as well, the traditions of Isis and the fourth epagomenal day in later times.

4 The Latin designation is still utilized by Leo Depuydt in "On the Consistency of the Wandering Year as Backbone of Egyptian Chronology," *JARCE* 32 (1995): 43–58. I see no reason why to apply this term, a late one historically speaking, to what in essence was an invention of the state. The German "bürgerlicher Kalender" aptly expresses the new system as well as the English "Civil Calendar." Cf. as well Depuydt's study "Calendars and Years in Ancient Egypt," in John M. Steele, ed., *Calendars and Years: Astronomy and Time in the Ancient Near East* (Oxford, 2007), pp. 35–81 — but add my review of the volume in *JAOS* 131 (2011): 311–13. One final point of historical importance: the implementation of the Egyptian Civil Calendar was caused by the state's desire to codify and make it uniform as much as possible. Granted that Egypt was then an archaic subsistence agricultural country. Nonetheless, the ruling class, thin though it was, needed to organize their kingdom with regularity and firmness; i.e., it desired, as all governments do, control. With this background analysis one can not understand why Depuydt returned to the *annus vagus* designation. However, the employ of different terms is not mere scholarly squabbling. All hinges upon a socio-historical sense of the past. A useful parallel might be drawn between the coining of the two *termini technici* — "Civil Calendar" and Civil Society." The present term "civil society," first employed by English political economists, gradually took over in the lands of Germany in the eighteenth century AD. The translation was "bürgerliche Gesellschaft." I ignore the later ambiguity inherent in the German term: Stephen Marcus, *Engels, Manchester, and the Working Class* (New York, 1974), pp. 133–34, note 5. One can blame Hegel.

5 The importance of Sothis is demonstrated by Krauss in *Sothis- und Monddaten*. But his argumentation is also a riposte to that of Richard Parker in *The Calendars of Ancient Egypt* (Chicago, 1950). I find that I cannot take seriously Patrick O'Mara, "Censorinus, the Sothic Cycle, and Calendar Year One in Ancient Egypt: The Epistemological Problem," *JNES* 62 (2003): 17–26. Krauss' last summary on this matter is "Astronomical Chronology," in Juan Antonio Belmonte and Mosalam Shaltout, eds., *In Search of Cosmic Order: Selected Essays on Egyptian Archaeoastronomy* (Cairo, 2009), pp. 135–54. This is the only seriously presented contribution in an otherwise uninspiring volume. Earlier, see his "Egyptian Calendars," *KBN* 3 (2008–9): 105–13. For a useful study of

role in Egyptian astronomy and religion, and since it is the star with the greatest visible magnitude, such a hypothesis met with little scholarly refutation for over one century of Egyptological research. Furthermore, Sothis was connected to the annual inundation, the commencement of New Year's Day, the rebirth of the land of Egypt, and the reinstitution of the powers of pharaoh, and thus such an interpretation made a great deal of sense. Then, in 1938, Otto Neugebauer challenged this opinion. He argued that it would have been simpler for the Egyptians to have averaged out the annual floods of the Nile, and hence they would have quickly seized upon the integer of 365 days.[6] Both his position and the one he overthrew set the commencement of the ancient Egyptian year in mid July.

Neugebauer carefully assessed the intellectual capabilities of the Egyptians from late prehistoric times through the Archaic Age of Dynasties I–II. One problem of modern Egyptological scholarship has always been the willful belief that pharaonic Egypt was far more exact, calendrically speaking, than the evidence suggests.[7] Neugebauer was wise to analyze Egyptian astronomical and calendrical development within their society and not ours. Furthermore, he repeatedly measured the economic, social, mathematical, and technological know-how of Egypt at specific points in time. Therefore, when Neugebauer sifted through the immense scholarly literature dealing with the original of the Civil Calendar, he derived a more reasonable justification for its origins.

It may be useful in this context to specify why the term Civil Calendar is appropriate. When Egypt was unified around 3100 BC, it was necessary for the incipient bureaucracy to enact a regularized system of governance. Essential for any large

---

New Year's Day in Egypt, see J. F. Borghouts, *Nieuwjaar in het Oude Egypte* (Leiden, 1986).

6 Parker, *The Calendars of Ancient Egypt*, pp. 51–52. The basic studies of Neugebauer are "Die Bedeutungslosigkeit der 'Sothisperiode' für die älteste ägyptische Chronologie," *Acta Orientalia* 17 (1939): 169–85, and "The Origin of the Egyptian Calendar," *JNES* 1 (1942): 396–403. Krauss modifies some of these presuppositions on pages 178–80 of his *Sothis- und Monddaten*. It was Robert Walter Sloley in "The Origin of the 365-Day Egyptian Calendar," *ASAE* 48 (1948): 261–65, who effectively refuted Neugebauer. The arguments presented here will be discussed in detail later.

7 I have repeated this contention over a period of years. Some scholars, bringing with them little mathematical and astronomical know-how, still prefer to apply a modern scientific approach to their study of ancient Egyptian calendrics. For example, see Christian Leitz, *Studien zur ägyptischen Astronomie* (Wiesbaden, 1989). Note the two important reviews in *OLZ* 87 (1992): 23–26 (Anthony Spalinger) and *BiOr* 49 (1992): 721–28 (Ronald Wells). An extreme example is that of Lynn E. Rose, *Sun, Moon, and Sothis: A Study of Calendar Reforms in Ancient Egypt* (Deerfield Beach, FL, 1999). Two critical reviews will be found in the *Bryn Mawr Classical Review* 2000.01.17 (online), 102 (Spalinger), and *JNES* 61 (2002): 311–15 (Wells).

state is the ability to control, especially internally but also externally, its political and military dispositions. Hence, we find very early in the origin of civilizations an array of standardized social patterns, among which we can signal court, capital, bureaucracy, juridical system, internal policing, and so forth. Concomitant with the state was the necessity of establishing a formal means of dating and bookkeeping. For this reason alone a good argument can be made for the establishment of the Civil Calendar not too long after the beginning of the First Dynasty. Hitherto, there had been only a lunar-based calendar, one subject to the vagaries of the moon.

According to Neugebauer, the Egyptians utilized their knowledge of the inundation of the Nile as the inspiration for a year of 365 days. Richard A. Parker demurred. In 1950, he devised a more detailed and systematic argument wherein the Civil Calendar arose from the preexistent lunar calendar by the method of approximation of the lunar cycle.[8] Accepting as fact the evidence for the earlier lunar calendar, he thought it was unnecessary to look to Sothis or the Nile inundation in order to construct the new annual schematic system; rather, by averaging the lunar year over a period of time, the 365-day cycle would surely have emerged. This argument had the advantage of avoiding any association with natural phenomena not directly connected with a "year." Parker's lunar hypothesis makes the reasonable assumption that calculations remained based upon the moon even with the elimination of a "real" lunar year. (According to Parker, following Ludwig Borchardt, the Egyptian lunar calendar must have operated with an intercalary month of 30 days occurring in the second, third, or fourth years.)

Parker's analysis had the salutary effect of accustoming Egyptologists to the origins of the Civil Calendar through recognition of the earlier lunar one. Moreover, since traces of lunar-based festivals can be found throughout pharaonic history and even during the Graeco-Roman Period, Parker's contention met with little opposition. Only recently have there been serious challenges to his position. In 1985, Rolf Krauss rejected the Parker hypothesis by placing the commencement of the Civil Calendar at one of the key dates of Sothis, 2738 BC, as determined astronomically from the equation of the first day of the year with the heliacal rising of Sothis.[9]

The connection to this star's reappearance in the eastern horizon after its period of invisibility makes sense calendrically, because the basic parameters of Egyptian calendrics and astronomy originally lay in the east. This applies to the heliacal rising

8 *Calendars*, pp. 51–56.

9 *Sothis- und Monddaten*, pp. 191–92. I have presented a developed argument concerned with the origin and the dating of the Civil Calendar in "The Beginning of the Civil Calendar," in Miroslav Bárta, Filip Coppens, and Jaromír Krejčí, eds., *Abusir and Saqqara in the Year 2010* (Prague, 2012), pp. 723-35. This study is covered in the subsequent discussion, "The Results of the Civil Calendar."

of Sirius as well as to the determination of the first lunar day.[10] The latter is an eastern phenomenon, for the last lunar day occurs when the final phase of the moon can still be seen in the east before it vanishes. Disappearance marks day 1 (Egyptian); its name derives from the old word meaning "new."[11] On day 2, the moon's crescent can be first seen in the west. Krauss also commented that with the introduction of the Civil Calendar not only were the Egyptians locked into a new cycle of the star Sothis, but also that this simplified year roughly coincided with the expansion of the sun cult in Dynasties II–III. He dated "the introduction of a sun calendar" to this time.

In many ways this last analysis seems the most reasonable. With Ronald Wells, Krauss looked for solar implications, and there is little doubt that an averaging of the sun's return to a specific point in its annual longitudinal circuit could easily have been determined by the Egyptians at the turn of the Fourth to the Third Millennium BC. It is just that Krauss further coordinates this solar calendar with Sothis.

Meanwhile, two studies by Wells offered a novel interpretation of the Civil Calendar originating in two predynastic lunar calendars and connected to the cosmology of Nut, the sky goddess.[12] Like Krauss, Wells maintained that it was far simpler for the Egyptians to average a solar year than to work with lunar months. But he felt that there was a dual origin for the 365-day period, one based on a lunar-solar calendar

---

10 The commencement of the Egyptian day is still controversial. For two conflicting arguments, see Krauss, "Was wäre, wenn der altägyptische Kalendertag mit Sonnenaufgang begonnen hätte?," *BSEG* 17 (1993): 63–71—mainly directed against the unsubstantiated suppositions of Christian Leitz; Ulrich Luft, "Rund um den Beginn des ägyptischen Tages," in Ernst Czerny et al., eds., *Timelines: Studies in Honour of Manfred Bietak* I (Leuven, Paris, and Dudley, 2006), pp. 207–15; and Spalinger, "Land Brightens, Day Begins," in Spalinger, *Five Views on Egypt* (Göttingen, 2006), pp. 51–85.

11 A fact not ignored by Parker, *Calendars*, p. 12 and note 38, citing Kurt Sethe, *Von Zahlen und Zahlworten bei den alten Ägyptern und was für andere Völker und Sprachen daraus zu lernen ist: Ein Beitrag zur Geschichte von Rechenkunst und Sprache* (Strasburg, 1916), p. 20. Antonio Loprieno, "Zahlwort," in Wolfgang Helck and Eberhard Otto, eds., *Lexikon der Ägyptologie* (Wiesbaden, 1986), col. 1308, independently saw the connection. Why "new" = 9? Because, in physical counting, the first *modus operandi*, a hand has four fingers and 4 × 2 = 8 (for two hands). Hence 9 begins something (a)new. Originally, thumbs were not considered to be fingers, as the English name of the appendage still indicates.

12 "Re and the Calendars," in Anthony Spalinger, ed., *Revolutions in Time: Studies in Ancient Egyptian Calendrics* (San Antonio, 1994), pp. 1–37, and "The Mythology of Nut and the Birth of Re," *SAK* 19 (1992): 305–21. Some of his argumentation has now been questioned by Joachim Quack in his unpublished Habilitationsschrift for the Freie Universität zu Berlin, *Untersuchungen zu den ägyptischen Dekanen und ihrer Rezeption in der griechisch-römischen Welt* (Berlin, 2002 submitted, accepted 2003), in press.

of Lower Egypt and a lunar-stellar one of Upper Egypt. Although somewhat complicated, this reconstruction has the advantage in linking up the winter solstice (*mswt R*ᶜ, the "Birth of Re" — the name of the twelfth Egyptian month) with the later Civil Calendar. Recent work by Alexandra von Lieven and Joachim Quack, however, tends to indicate that the "law of parsimony" should hold. Namely, that the extant data permit only a unilateral development within Egypt.[13]

It would be useful if the early First and Second Dynasty records contained any evidence for the use of the Civil Calendar. They do not. What we see instead is the gradual development of a supplementary bureaucratic system of calendrical notation: the regnal year. The first evidence for such notations occurs right at the beginning of the First Dynasty.[14] We can ascertain from various ivory labels, vase inscriptions, and tablets that each year of a king was given a specific name. Unfortunately, it is

---

13 Quack's major work is cited in the previous note. Alexandra von Lieven's study is *Grundriss des Laufes der Sterne: Das sogenannte Nutbuch* (Copenhagen, 2007).

14 An extremely useful presentation of this material will be found in Donald B. Redford's *Pharaonic King-Lists, Annals and Day Books* (Mississauga, 1986), pp. 86–90. The key publications include Heinrich Schäfer, "Ein Bruchstück altägyptischer Annalen," *Abhandlungen der Königlichen Preussischen Akademie der Wissenschaften zu Berlin: Phil.-hist. Klasse, Anhang Abhandlungen* 1 (1902); Henri Gauthier, "Quatre nouveaux fragments de la pierre de Palerme," *Le Musée Égyptien* 3 (1915): 29–53; H. M. Stewart, *Egyptian Stelae, Reliefs and Paintings from the Petrie Collection* (Warminster, 1979), p. 6 (UC 15508) and plate 3.1; with J. L. de Cenival, "Un nouveau fragment de la pierre de Palerme," *BSFE* 44 (1965): 13–17. The new compendium of Toby A. H. Wilkinson, *Royal Annals of Ancient Egypt: The Palermo Stone and its Associated Fragments* (London and New York, 2000) replaces all of these earlier studies. Nonetheless, the discussion of the parameters of regnal year dating have not altered much — or indeed significantly — since the publication of Sir Alan Gardiner's "Regnal Years and Civil Calendar in Pharaonic Egypt," *JEA* 31 (1945): 11–28. The only major improvement concerns the difficulty in ascertaining the presence of the biennial counts and their disappearance, for which see now Miroslav Verner, "Contemporaneous Evidence for the Relative Chronology of Dyns. 4 and 5," in Erik Hornung, Rolf Krauss, and David A. Warburton eds., *Ancient Egyptian Chronology* (Leiden and Boston, 2006), pp. 124–43; with Michel Baud, "The Relative Chronology of Dynasties 6 and 8," pp. 144–58 in the same volume. Pages 19–25 in the same edited volume also cover the "Royal Annals," almost as an afterthought. The first evidence for the regnal year sign is in the reign of Den of the First Dynasty: Kathryn A. Bard, "The Emergence of the Egyptian State," in Ian Shaw, ed., *The Oxford History of Ancient Egypt* (Oxford, 2000), p. 81. Verner's study, "The System of Dating in the Old Kingdom," in Hana Vymazalová and Miroslav Bárta, eds., *Chronology and Archaeology in Ancient Egypt: The Third Millennium B.C.* (Prague, 2009), pp. 23–43, remains the basic study of regnal years at this time. John S. Nolan's contribution in the same volume on pages 44–60, "Lunar Intercalations and 'Cattle Counts' during the Old Kingdom: The Hebsed in Context," resembles Philippians 4.7.

impossible to determine whether these years were lunar or whether they reflect the Civil Calendar. It is clear, nonetheless, that at the beginning of the Third Dynasty there was a solidification of the state, but the records are mute with regard to this calendar. Only when we get to the Fourth Dynasty is civil reckoning chronicled in contemporary documents as well as in the later Palermo Stone (and its Cairo fragments).[15] It is unknown whether this has to do with the rise of the cult of the sun, exemplified by pharaohs' relationship to the sun god Re, his father, and the sky goddess Nut, his mother.

Those First Dynasty labels, described by Donald Redford, show a progression in organization and a gradual standardization throughout that period.[16] By the Second Dynasty, the basic method of dating had been established. The subsequent replacement of specific names for the years through the rise of a biennial cattle census then took place. Records commenced with "year of count X" or "year after count X." At the end of the reign of Pepi II it appears that the biennial system was abandoned and that all regnal year dates referred to an individual year. The regnal year system enabled the Egyptians to count each pharaoh's reign according to a regular system of calendrics; namely, the Civil Calendar. In the Palermo Stone we find the scheme fully operational for the first time under the Fourth Dynasty ruler Shepseskaf; earlier entries are inscrutable.[17]

That the Egyptian civil months were based upon earlier lunar ones is easy to see. Even though for the most part we have to reply upon New Kingdom data, it is clear that each civil month had a name as well as a number.[18] Thoth, for example, was the first civil month, although the deity had replaced the earlier civil designation of *Tḫj*.[19] Then came Phaophi, again indicating a New Kingdom designation which had replaced *Mnḫt*. (The association to "clothing" surely reflects a calendrically-specified offering or offerings.) We can reconstruct the original names of the Civil Calendar

---

15 Michel Baud and Vassil Dobrev, "De nouvelles annales de l'Ancien Empire égyptien. Un 'Pierre de Palerme' pour la VI[e] Dynastie," *BIFAO* 95 (1995): 23–63; and add their "Le verso des annals de la VI[e] Dynastie," *BIFAO* 97 (1997): 35–42.

16 See note 14, above, for the data.

17 Gardiner, "Regnal Years and Civil Calendar in Pharaonic Egypt," p. 12; note as well Ludwig Borchardt, *Die Annalen und die zeitliche Festlegung des alten Reiches der Ägyptischen Geschichte* (Berlin, 1917).

18 A handy summary is presented by Parker in his *Calendars*, pp. 11–13. Henceforth, all month names will be capitalized, whereas I have not done this with the feasts.

19 *Calendars*, p. 31. A controversial discussion of the lunar-civil situation of the months may be seen in Depuydt, *Civil Calendar and Lunar Calendar in Ancient Egypt* (Leuven, 1997). His lengthy study can be countered. See my "The Beginning of the Civil Calendar," and the subsequent discussion "The Results of the Civil Calendar," but already note the review of Christian Leitz in *BiOr* 57 (2000): 75–81.

months, although over time new designations replaced these ones. For example, the aforementioned Phaophi indicates the importance of the New Kingdom Opet festival. Similarly, the seventh civil month, Phamenoth, reflects the deification of Amunhotep I. Scattered Middle Kingdom data enable us to turn the pages back, so to speak.[20]

According to Parker, the earlier lunar calendar contained the twelve expected lunar months plus one intercalary one, which he named Thoth, the moon god. However, his contention was mainly based upon two-dimensional representations of clepsydra (water clocks) in which the spigot through which the water trickled away bears the figure of Thoth.[21] But in these representations, as well as on the few exemplars of actual clepsydra, such as the famous one of Amunhotep III,[22] the moon god was positioned at the exact place between the end of the civil year (month twelve) and the beginning of the first civil month (*Tḫj*). Therefore, the association of an ever-present intercalary month called Thoth must be dismissed.[23]

---

20 An earlier attempt of mine to reconstruct the primitive calendrical situation in Egypt may be found in "Notes on Ancient Egyptian Calendars," *Or* 64 (1995): 17–32. I no longer subscribe to some parts of the analysis.

21 For the importance of water clocks for night length rations and the assumed "errors" in ancient Egyptian clepsydra, see John Fermor, "Perceived Night Length Ratios in Ancient Egypt," *Vistas in Astronomy* 36 (1993): 363–73.

22 See my "Month Representations," *CdÉ* 70 (1995): 110–22. Add Charlotte R. Long, *The Twelve Gods of Greece and Rome* (Leiden, New York, Copenhagen, and Cologne, 1987), pp. 147–51.

23 Owing to the invention and development of the water clock in Mesopotamia centuries earlier than its presence in Egypt — Neugebauer, "The Water Clock in Babylonian Astronomy," *Isis* 37 [1947]: 37–43 — I argued that at least the concept of measuring time by water was borrowed by the Egyptians at the beginning of the Eighteenth Dynasty in "Some Times," *RdE* 47 (1996): 67–77. Cf. Brian Cotterell, F. P. Dickson, and Johann Kamminga, "Ancient Egyptian Water-clocks: A Reappraisal," *Journal of Archaeological Science* 13 (1986): 37, who also have raised doubts concerning the assumption that Amnemhet "invented" the water clock for Egypt. A useful summary is presented by Neugebauer and Parker in *Egyptian Astronomical Texts I: The Early Decans* (Providence and London, 1960), pp. 118–19. All of this material has been discussed by Pierangelo Mengoli in his compendium *Astronomica Egizia* (Budrio, 2006). Nonetheless, the following three published articles may be consulted with profit: earlier writings can be found in his "La clessidra egizia del Museo Barracco," *VO* 6 (1986): 193–209, "Some Considerations of Egyptian Star-Clocks," *Archiv der Geschichte der Naturwissenschaften* 22–24 (1988): 1127–50, and "La clessidra di Karnak: L'orologio ad acqua di Amenophis III," *OA* 28 (1989): 227–71. Add these studies: Ludolf von Mackensen, "Neue Ergebnisse zur ägyptischen Zeitmessung," *Alte Uhren* 1 (1978): 13–18; Josef Dorner, "Tagesuhr-Nachtuhr: Ein Beitrag zur altägyptischen Zeitmessung," *Jahreshefte des Österreichischen Archälogischen Instituts in Wien, Beiblatt* 65 (1985):

Ever since Heinrich Brugsch, Egyptologists have noted the association of these names with various festivals. For example, in the first month there occurred the fixed feasts of *w3gj*, Thoth, and *tḫj*, held on days 18, 19, and 20, respectively.[24] Hence, it is not hard to see why the original civil name for month one was *Tḫj* and later Thoth. Complications arise when we attempt to ascertain the causes for the following:

a. Month three is called Hathor, yet the celebration for this goddess occurs in the next month, usually on day one of civil month four.
b. Month four is called Choiak (*K3-ḥr-k3*), effectively named after a major event originally enacted on days 25–26, but later expanded up to the last day of the civil fourth month. Soon the feast Choiak was connected with the festival of Nehebkau (*nḥb-k3w*), which took place on day one of the fifth civil month. The entire series of rites and performances was correlated to the death and resurrection of the god Osiris, with the beginning of the second season marking an effective break.
c. The festival of Renutet (*rnwtt*) is placed to day one of the ninth civil month, yet the name of the preceding civil month is the same.
d. The two middle months were referred to as "Fire; Heat," *Rkḥ*. Parker attempted an explanation for the sixth and seventh months as owing

---

27–37; Maria G. Fineis, "Astronomische Bestimmung der geographischen Breite aus den Markierung einer ägyptischen Wasseruhr," *Jahreshefte des Österreichischen Archäologischen Instituts in Wien* 55 (1987): 9–11 and 18–19; Sylvia Couchoud, "Calcul d'un horloge à eau," *BSEG* 12 (1988): 25–34; Willibad Henne, "Überlegungen zur altägyptischen Wasseruhr," *GM* 115 (1990): 45–56; and Philippe Derchain, "La clepsydre de la dernière nuit," *GM* 210 (2006): 19–21. The Amunhotep III water clock is nicely discussed in "La clepsydre de Karnak," in C. Desroches-Noblecourt, *Ramsès le grand* (Paris: Galeries Nationales de Paris, 1976), pp. 139–49.

24 These are the basic studies associated with the three key feasts listed here. *W3gj*: Luft, "The Date of the *W3gy* Feast: Considerations on the Chronology of the Old Kingdom," in Spalinger, ed., *Revolutions in Time*, pp. 39–44; Krauss, "Wenn und Aber: Das Wag-Fest und die Chronologie des Alten Reiches," *GM* 162 (1998): 43–63—attempting a refutation of Luft; and Hana Vymazlová, "Some Remarks on the *w3g*-Festival in the Papyrus Archive of Raneferef," in Vymazlová and Bárta, eds., *Chronology and Archaeology in Ancient Egypt*, pp. 137–43. Krauss, "Ein Versuch zur Chronologie des späten Alten Reiches im Anschluss an die Monddaten im Neferirkare-Archiv," *Or* 77 (2008): 377–85, is a useful attempt to set into place a "new remote" chronology. Thoth: Spalinger, "Thoth and the Calendars," in Spalinger, ed., *Revolutions in Time*, pp. 45–60. *Tḫj*: Spalinger, "A Chronological Analysis of the Feast of *Tḫy*," *SAK* 20 (1993): 289–304. In general, see Spalinger, "Egyptian Festival Dating and the Moon," in John M. Steele and Annette Imhausen, eds., *Under One Sky: Astronomy and Mathematics in the Ancient Near East* (Münster, 2002), pp. 379–403.

to the need for warmth in the middle of the Egyptian year.[25] Whereas this appears possible, the question remains as to the origin of the lunar year and whether that starting point coincided with the later civil one.

e. The twelfth and last civil month was called *Wp rnpt*, or the time when "the year is opened." Surely, Egyptologists have argued, this ought to be the name of the beginning of the year and not its end. In identical fashion, the later designation for the twelfth month, Mesore, the "Birth of Re" (the sun god), ought to have been equally employed for the opening month in a year.

Another consideration is the ecological connection.[26] For example, the feast of *rnwtt* at the beginning of civil month nine, which was named for the goddess of grain, followed the "middle months" of "heat." It was not unprecedented, as Martin Nilsson saw, to have the central portion of a year consisting of two months possessing the same name.[27] Apparently, in the prehistoric period it was deemed convenient to subdivide a rather large period of time into two roughly equal portions: "Great Heat" (month six in the later civil calendar) and "Small Heat" (civil month seven). When the Civil Calendar designation of *Rnwtt* was taken over, it merely retained a name that had been in use since the calendar was lunar. In other words, the newer calendar followed for the most part the earlier lunar-based one with respect to the names of the month, even though there was a reordering of the arrangement, wherein the name for month I now became that of month XII, and so forth. Perhaps of equal significance is the strong possibility that in the earlier lunar calendar, the middle was originally designated "heat." In this case, the later movement of lunar month names to civil ones quite possibly preserved the core or center of the first system and thus incorporated it, rather easily, into the "rational" second.

---

25 Parker, *Calendars*, p. 46.

26 Spalinger, "Notes on Ancient Egyptian Calendars." In his "Re and the Calendars," Wells, on the other hand, proceeds from the assumption that the name of the twelfth Egyptian month, *Mswt Rᶜ*, indicates the winter solstice. He then proceeds to locate two predynastic lunar calendars: one in Lower Egypt which was connected with solar worship and therefore based on the winter solstice; the second, in Upper Egypt, was then governed, in contrast to the former, by the starting point of the heliacal rising of Sothis, *prt Spdt*.

27 Martin Nilsson, *Primitive Time Reckoning* (Lund, 1920), pp. 224–25. This is an excellent empirically-based study, but to use the phrase of Walther Wolf in his review of William Stevenson Smith's *The Art and Architecture of Ancient Egypt* (Harmondsworth, 1958), it is a classic example of "reine Dingswissenschaft": *OLZ* 56 (1961): 18; cf. Marianne Eaton-Krauss in her review of the later "third edition" of William Stevenson Smith, *The Art and Architecture of Ancient Egypt* (New Haven, 1998): *OLZ* 95 (2000): 399.

As for the seasons, only the Civil Calendar reveals their names and time spans. There is no indication how the Egyptians considered them to be before the invention of the Civil Calendar. Each season comprised fourth months, identified with phases in the subsistence economy of Egypt. Season one, called "inundation," *ꜣḫt*, was a logical designation for the beginning of the year. Once more we can surmise how regular and systematic the second Egyptian calendar was. In many ways the year was actually divided into four portions, if not more. There is no evidence that any original Egyptian season consisted of four months. Later, each civil season consisted of four months. The second, also agriculturally ordered, was "emergence" or *prt*, whereas the final one was "harvest," *šmw*. But earlier, one might wish to argue—with a high degree of speculation to be sure—for four seasons: inundation, rebirth, grain-emergence, and harvest.

The yearly division reflects the domestication of grain, but such an exact arithmetical organization of four months apiece seems highly schematic. Moreover, the order of the civil seasons implies a commencement of the year in mid-July, a fact which justifies Krauss' arguments for the origins of the Civil Calendar on the date when the star Sothis reappeared (Julian 19 July) and the inundation began.[28]

The basis was the same for naming the seasons as the months: ecology and the anticipatory procedure of farming. We have already noted the names for the middle months, six and seven. Another indication was *Šf bdt* for the "swelling of the grain," emmer wheat. The expansion of the wheat during month five, coincidentally the first month of season two, fits the normal seasonal flow in Egypt. Later, religious association of offering replaced earlier ones. This must have been due to a concatenation of alterations, not the least was the growing development of an archaic state whose conception more and more diverged from the earlier and simpler agriculturally ordered universe. That it to say, the subsistence agricultural metaphors and *termini technici* of prehistoric Egypt lost their significance when Egypt turned to a highly developed urban civilization.[29] For example, *ꜥꜣbt* > *Tꜣ-ꜥꜣbt*, Tybi, which Gardiner labeled "banquet," came to replace the earlier name for the fifth civil month, the agrarian-based *Šf bdt*. In similar fashion the term Payni, indicating the famous New Kingdom Valley Feast (*ḥb tꜣ jnt*), became the name for civil month ten, having

28 Recent studies on Sothis include "The Heliacal Rising of Sirius," by Teije de Jong, and "Egyptian Sirius/Sothic Dates and the Question of the Sirius Based Lunar Calendar," in Hornung, Krauss, and Warburton, eds., *Ancient Egyptian Chronology*, pp. 432–38 and 439–57, respectively. In general, see Krauss, *Sothis- und Monddaten*, pp. 189–92. But primary scientific naked-eye observational studies are lacking in these analyses.

29 For the following example, see Gardiner, "Thutmosis III Returns Thanks to Amun," *JEA* 38 (1952): 23; and Jaroslav Černý, "The Origin of the Name of the Month Tybi," *ASAE* 43 (1943): 173–81, "Philological and Etymological Notes," *ASAE* 51 (1951): 441–42, and "Some Coptic Etymologies III," *BIFAO* 57 (1958): 206–8.

ousted *Ḫntj-ḫtj*.[30] To repeat ourselves for the moment, two additional cases where cultic paraphernalia or offerings served as the original name of a lunar month may be proffered: Epiphi and *Mnḫt*, "Clothing,"[31] the last of which was eventually replaced by Phaophi, thus now referring to the great New Kingdom Theban celebration of Opet. Mechir, also connected to cultic services, took over from the even more archaic seasonal designation of *rkḥ wr* for civil month six.

---

30 Cf. Černý, "Philological and Etymological Notes," *ASAE* 51 (1951): 441–42.

31 Luft, *Die chronologische Fixierung des ägyptischen Mittleren Reiches nach dem Tempelarchiv von Illahun* (Vienna, 1992). Krauss had some difficulties in accepting part of the conclusions presented ("Fällt im Illahun-Archiv der 15. Mondmonatstag auf den 16. Mondmonatstag?," *GM* 138 [1994]: 81–92), but see the later reaction of Luft, "'Weil ... nicht sein kann, was nicht sein darf'," *GM* 141 (1994): 109–11. A further attempt by Krauss to provide an effective counter to the study of Luft can be read in his review of Luft's *Die chronologische Fixierung* in *OLZ* 89 (1994): 6–18. The recent analysis of the same material by Rita Gautschy, "Lunar and Sothic Data from the Archive of el-Lahun Revisited: Chronology of the Middle Kingdom," in Maarten Horn et al., eds., *Current Research in Egyptology 2010: Proceedings of the Eleventh Annual Symposium* (Oxford and Oakville, 2011), pp. 53–56, needs additional commentary. See the study of Ronald Wells at the end of this chapter in the Appendix. In her "Monddaten aus dem Archiv von Illahun: Chronologie des Mittleren Reiches," *ZÄS* 138 (2011): 1–19, Gautschy refers to B. D. Yallop, *A Method for Predicting the First Sighting of the New Crescent Moon*. (*HM Nautical Almanac Office NAO Technical Note* 69) (Chilton, June 1997 and updated April 1998). He went through the Bradley Schaefer material and improved the observations which had errors in them. In addition, Yallop re-derived an algorithm for predicting the first crescent moon based on Schaefer's approach. On page 5 he presents this important methodological approach: "As Schaefer points out, using different extinction coefficients I find an enormous difference between a site at high altitude with a clear dry atmosphere, and one at sea level with a humid and dusty atmosphere, so my empirical approach must be confined to a specific type of site where altitudes above sea level and extinction factors are confined to within narrow limits." But how one uses it is still a problem. The passage just cited emphasizes that. In the concluding remarks of his "The Role of Astronomical Techniques in Ancient Egyptian Chronology," p. 470, Wells stated: "This being the case, is it worth asking whether there are any conditions in which the technique applied to Egypt might be useful for absolute chronology? That is, would it be worthwhile re-investigating, for example, the Illahun sequence using the more rigorous algorithm provided by Schaefer. The answer has to be no, that none of the Egyptian lunar dates offer any promise of yielding an absolute date for two reasons: (1) the likelihood that any observed sequence of multiple month lengths contains at least one error, but maybe more, invalidates its use; and (2) the large number of consecutive month lengths, given perfect observations and perfect computations of past events, required for statistical validity far exceeds the available Egyptian record." Gautschy also talks about computing the last crescent—this is quite different than

## B. FEASTS: ORIGINAL AND LATER

The arrangement of Civil Calendar month naming, the overt artificiality of time-keeping, and the unnatural regularization of the three seasons together managed to obscure the original lunar-based system of prehistoric times and probably the Archaic Age. The most effective reform must have been the displacement between feasts and the names of the civil months. Some but not all of the month names can be linked to the corresponding feast at the commencement of the following month. What, then, can we conclude about the original lunar system? Unfortunately, many Egyptologists fail to see that this query is directly related to the establishment of the Civil Calendar on the first day of its inception. To take a key example: civil month twelve, *Wp rnpt*, has the same designation as New Year's Day in the same calendar. Can we assume that there was a *décalage* at the inauguration of the Civil Calendar—let us say, following Krauss, about 2738 BC—when day 1 of that new calendrical system was fixed at the heliacal rising of Sothis?[32] If so, then the oddity of month designations preceding the actual feasts (usually held or begun also on day one of the following civil month) can make some sense.

Preceding the enactment of the Civil Calendar there was an earlier lunar-based one. Given that the lunar-solar epact is plus eleven days or minus nineteen days (the difference between 354 days, a standard lunar year in integers, and 365 days, a standard solar year in integers), the following may be hypothesized:

1. At the inauguration of the Civil Calendar the preexistent year would have ended before 365 days.
2. As there is no indication that the Egyptians had an intercalary month at this time, we can ignore the possibility of that year's containing thirteen lunar months. If there is doubt, then let us simply set the end of the last official state lunar year on day 354.
3. In the earlier system, New Year's Day would occur on day 355 by mere counting. The month would be the first one of the year and so be designated by *Wp rnpt*.

---

Yallop's algorithm and is covered by Wells in his detailed note at the end of this chapter. In passing, it should be observed that Krauss rebukes Wells' positions in "Ronald A. Wells on Astronomical Techniques in Ancient Egyptian Chronology," *DE* 57 (2003): 51–56. Yet see the Appendix to this chapter.

32 Krauss, *Sothis- und Monddaten*, p. 192. In "The Beginning of the Civil Calendar," I followed Oleg D. Berlev, who placed the first Sothic cycle as commencing in the reign of Djoser: "Precisely Two Sothic Periods Between Year 18 of King *Sénw*/Tosorthros and Year 2 of Pharaoh Antoninus Pius," in O. I. Pavlova, ed., *Ancient Egypt: Language, Culture, Consciousness* (Moscow, 1999), pp. 42–62.

4. If the lunar year were to have continued for official bureaucratic state use, then lunar month two, *Tẖj*, would have taken place around day 385 (354 + 30 or 354 + 29).
5. The festival associated with lunar month one was *wp rnpt*, and it would have begun the month.
6. Equally, the festival associated with lunar month two was *tẖj*, and from later civil sources of the historic era it is known to have been civilly fixed to the first month of *ꜣẖt* (inundation), day 20.
7. Now there are precisely thirty-one days between the last day of the outgoing year and the festival of *tẖj*.
8. In order that the newer designation for the first civil month be *Tẖj*, when the civil dating of the same-named festival was the twentieth, this must imply a difference of 29-9, allowing the length of a lunar month in this case to be twenty-nine days.
9. Therefore, when the Civil Calendar came into being, at least nine days had passed by in the original lunar year.

The festival of *wp rnpt*, which should have occurred on day 355, was now set on day 366, eleven days later. The festival of *tẖj* is equally determined by this epact. It was assigned to day 20 of civil month one; but taking into consideration the lunar-solar difference, it is actually *nineteen* days later than civil *wp rnpt*.[33] So far there is relatively clear sailing. The crucial issue, one that concerns the exact date of the first day of the first civil month, has been discussed in another forum and its implications will be drawn in the following discussion. But for the moment it is easy to see why the later civil feast of Thoth was placed on day 19 of the first civil month: it simply represents the same epact of eleven days minus or nineteen plus.[34] To compound this analysis, the sorrowful celebration of *wꜣgj*, also associated with Osiris, occurred on day 18 of the first month in the same calendar. Previously, *wꜣgj* was set within the first lunar month on the 18th day.[35] The diagram in figure 1 will help to explicate the situation.

Therefore, it is reasonable to conclude that the beginning, or New Year's Day, of civil month one at the inception of the new calendar was actually later in the year than the commencement of the contemporary lunar year.[36] In addition, the sym-

33 Spalinger, "A Chronological Analysis of the Feast of *Tẖy*."

34 Spalinger, "Thoth and the Calendars."

35 Luft, "The Date of the *Wꜣgy* Feast."

36 Of course, this analysis is schematic. However, there is no way that the new Civil Calendar could be reconciled with the older lunar one unless a relatively fixed point was marked. For Krauss and Parker, this was the heliacal rising of Sothis, even though the two disagreed with respect to the original ancient determination for the 365-day year. I follow, to some degree, the argument of Krauss in "The Beginning of the

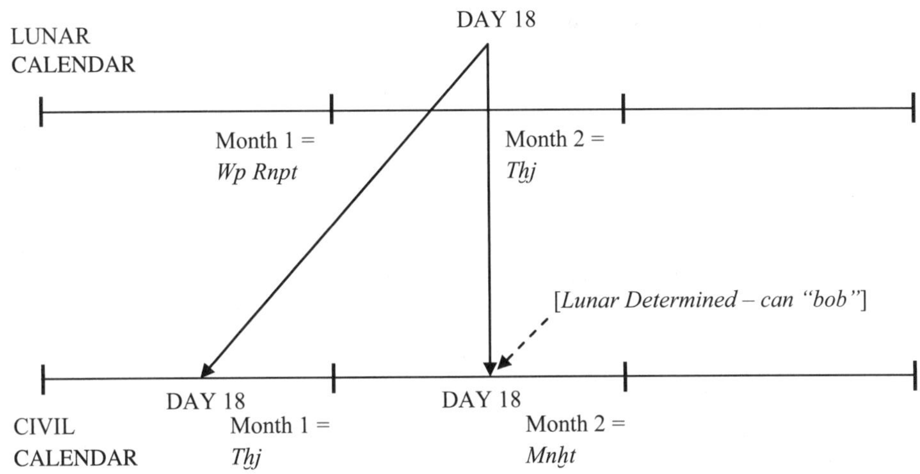


Fig. 1 The Feast of *W3gj*.

metry of the two religious feasts of *w3gj* and *tḫj* becomes apparent. This situation is inexorably linked to two key extant data that allow us to reconstruct this time, and the implications will be drawn even more tightly in the following discussion.

The symmetry of the two festivals *wp rnpt* and *tḫj* becomes apparent. Nineteen days separate the civil *tḫj* from the new civil *wp rnpt*; eleven days remain in the first month. Equally the same number of days separates the old lunar *wp rnpt* from the new civil one. The cause is simple: the new or incoming last civil month retained the name of the older, but chronologically earlier, lunar month one, whereas the new or first civil month, occurring later, was called *Tḫj*. To put it another way: the outgoing lunar calendar had its first month of *Wp rnpt* earlier than the first civil month.

A proposal by Leo Depuydt set out a very detailed and complicated explanation.[37] Yet, with Krauss, I believe that the Civil Calendar reflects the growing emphasis upon the sun as the determinate of the year. Whether or not we can verify that it occurred concomitantly with the rise of the solar cult is another matter. Parker, for example, claimed with underlined emphasis that "The civil calendar of 365 days was not tied to Sothis at its introduction but was tied rather to some yearly occurrence which was variable, so that the gradual shift forward of the civil calendar would not be immediately variable."[38] This hypothesis can be effectively challenged and a new paradigm proposed. Of greater importance is the underlying but not explicitly

Civil Calendar," although with new information. The analysis of Berlev indicates that this occurred in the reign of Djoser, a time period which I find questionable but not impossible.

37 *Civil Calendar and Lunar Calendar in Ancient Egypt*.

38 *Calendars*, p. 52. Parker underlined this passage for emphasis.

stated assumption of Parker. Namely, that there needed to be some regulation in the arrangement of the newer calendar. For Parker, this meant setting up the unwieldy development from lunar calendar > Civil Calendar > second lunar calendar. None of this is necessary.

In order to reconcile the gradual shift between the newly established Civil Calendar and its earlier lunar counterpart, Parker theorized this "second lunar calendar" whose "sole purpose would be to provide for the civil year the same sort of dualistic setup which had obtained when the civil year was first inaugurated."[39] Why not represent the eventual system of Egyptian calendrics as simply as possible: original lunar calendar > later Civil Calendar? Support for the second argument has now become overwhelming and is partly dependent upon the presence of moveable or variable lunar-based feasts recorded in the Civil Calendar. This, too, shall be covered in more detail later.

Contrary to some scholarly opinion, in the festival calendars there are actually few specific religious events set upon a day determined by the moon. The important lunar days that have been used in chronological reckoning refer primarily to one specific day — the first, the sixth, the fifteenth — all of which were especially important. If additional information is given, then the text refers to a festival or to an interval of time in which a major religious celebration was taking place. For example, at Medinet Habu, the southern exterior wall of Ramesses III's mortuary temple was inscribed with a detailed list of the regular feasts that were to be observed.[40] There are few of those events that were determined by the moon: the Feast of the Valley (List 3), the age-old Min feast (List 66), an unknown one (List 63), and a processional feast for

39 Ibid., p. 54.

40 *KRI* V 115–84; with Sherif el-Sabban, *Temple Festival Calendars of Ancient Egypt* (Liverpool, 2000), chapter 5. The study depends heavily upon Helck's major analysis, *Materialien zur Wirtschaftsgeschichte des Neuen Reiches* III. Akademie der Wissenschaften und der Literatur, Abhandlungen der Geistes- und Sozialwissenschaftlichen Klasse 1962.2 (Wiesbaden, 1962). This, however, is not given proper recognition in the volume. For example, Helck solved all of the complicated arithmetic in the Medinet Habu calendar, including which food items were measured by the quadruple hekat (oipe) and which ones by the sack. El-Sabban appears to have overlooked most of the work done in Helck's *Materialien*, since he interprets the grain measures as hekats. This is inaccurate. Keep in mind that the Medinet Habu calendar is actually copied from the mainly lost original (via use of the master endowment list) of Ramesses II. See Charles F. Nims, "Ramesseum Sources of Medinet Habu Reliefs," in Janet H. Johnson and Edward F. Wente, eds., *Studies in Honor of George R. Hughes: January 12, 1977* (Chicago, 1976), pp. 169–75; and Ben J. J. Haring, *Divine Households: Administrative and Economic Aspects of the New Kingdom Royal Memorial Temples in Western Thebes* (Leiden, 1997), passim, especially chapters 2–3.

Amun (List 67). Two are set on the first lunar day (*psḏntjw*).[41] A second reference to Min—"It is celebrated on the going forth of the 'Protector of the Moon'"—appears to refer to the first day of a lunar month when the moon is completely invisible.[42] Moreover, note the absence of the moveable *wꜣgj*; only the civilly based one is mentioned (List 25).

There is a striking difference between these few annual festivals and those recorded in the Illahun and Kahun papyri, which have been so well-assembled and discussed in Ulrich Luft's compendium.[43] The number of lunar feasts has diminished greatly. Similarly, the Eighteenth Dynasty festival calendars virtually ignore the existence of any religious celebration determined by the cycle of the moon. This discrepancy goes far in explaining the triumph of the Civil Calendar in the religious sphere of the Egyptians. Once the new system was put into place, all parts of the Egyptian administrative-accounting sectors adopted the regularized and standardized means of computing time.

In the festival calendar of Ramesses III, the main difference between feasts is not the expected one of civil versus lunar. In that document, the contrast is made between celebrations occurring more than once in the civil year and those occurring only once. The former are given the specific heading of "festivals of heaven" (*ḥbw nw pt*) and the latter are referred to as "seasonal festivals" (*tp trw*).[44] Yet, even the translation of those two terms fails to do justice to the ancients' thought. We find, for example, the festival of the heliacal rising of Sothis (List 23) at Medinet Habu listed under the *tp trw*. A quick survey of the material reveals that this separation has nothing to do with earth-based events —let us say human created ones—as opposed to astronomical or heaven-imposed ones. Rather, the dichotomy is founded upon regularity (more than once) versus singularity (once). Hence, we find the king's accession festival (List 21) logically placed under the *tp trw*, as it naturally occurred but once in the civil year. On the other hand, the *ḥbw nw pt* refer solely to various events in the monthly cycle of the moon.[45] They are organized as follows:

---

41 Readers should, I feel, take into consideration the remarks of Wells in his review of Rose, *Sun, Moon, and Sothis*: *JNES* 61 (2002): 313–14, wherein the difficulties in determining Egyptian *psḏntjw* are specified.

42 Parker, in a review of Siegfried Schott, *Altägyptische Festdaten*. Akademie der Wissenschaften und der Literatur, Abhandlungen der Geistes- und Sozialwissenschaftlichen Klasse 1950.10 (Wiesbaden, 1950), *BiOr* 9 (1952): 102.

43 *Die chronologische Fixierung des ägyptischen Mittleren Reiches.*

44 See chapter 1 of Spalinger, *The Private Feast Lists of Ancient Egypt* (Wiesbaden, 1996).

45 Ibid. Compare the breezy and amateurish approach of Byron E. Shafer, "Temples, Priests, and Rituals: An Overview," in Byron E. Schafer, ed., *Temples of Ancient Egypt* (Ithaca, 1997), p. 25, wherein this false statement is presented: "Some festivals, called *ḥbw nw pt*, were celebrated according to the astronomically determined sidereal calendar or according to the phases of the moon." Note 211 (on page 256) to this incorrect perception merely adds to the confusion.

a. Day 29: *ʿḥʿ...* = "lifetime [of the moon]"?; the end of a lunar month?
b. Day 30: "day of going-forth of Min," whether or not this came to pass[46]
c. Day 1: *psḏntjw*, "new day"
d. Day 2: *ȝbd*, "moon day"[47]
e. Day 4: *prt sm*, "day of going-forth of the *sm* priest"
f. Day 6: *snt*, "sixth day"
g. Day 10: *sjf*, the first decade
h. Day 15: [*tp*] *smdt*, "half month day."

There are the major lunar days with the first one, *psḏntjw*, being the most crucial, although the "mid-month" was very significant as well. Day 6, though important, ranked well behind these two. The Medinet Habu list, however, commences not with the first lunar day, which means something akin to "the new one," but rather the twenty-ninth. Reflected here must be an older way of approaching the days, whereas the outgoing one of the month might be the twenty-ninth or thirtieth. In that, a synodic lunar month averages 29.5309 days. Because the moon god is conceived on *psḏntjw*, as Parker noted, he must therefore be born on *ȝbd*. Impregnation definitely precedes *psḏntjw*. Hence, the attribution of the erect phallus of the deity Min and his "going forth" fits neatly into the ancient conception of the procreation of lunar months, one after the other. Finally, old age begins with day 15, (*tp*) *smdt*, which is logical, but no other lunar day is listed.

With regard to the "festivals of heaven," are their assigned days given only according to the moon? Second, they begin with the end of a phase rather than the beginning. Third, how are they determined? I.e., are there eight events based upon the lunar calendar or, rather, do they merely reflect the presence of key dates originally set within the lunar calendar (or month) but not pertaining to the Civil Calendar (or month)?

The first query is the most important one to resolve, since the number of purely lunar determined feasts diminished by the New Kingdom. Furthermore, many of those events were detected by eye rather than by the use of any lunar calendar. Thus

46 Cf. Harco Willems, *The Coffin of Heqata (Cairo JdE 36418): A Case Study of Egyptian Funerary Culture of the Early Middle Kingdom* (Leuven, 1996), who presents a confused situation on page 355 with note 2190. The deity *Nhs* was originally associated with the thirtieth lunar day; subsequently, he was carried over to the thirtieth civil day. What could be simpler?

47 This is a very difficult root to analyze. The root should be *lbd* or something similar. For the very simple Egyptian word *jʿḥw*, see Otto Rössler, *Gesammelte Schriften zur Semitohamitistik*, eds. Thomas Schneider and Oskar Kaelin (Münster, 2001), p. 576. Add Ludwig D. Morenz, "Der Mond. Über Königsnamen, Visuelle Poesie, Metaphorik und Zeitgeist in der Mitte des 2. Jt. v. Chr.," *BSEG* 24 (2002–3): 121–30.

the occurrence of a celebration taking place on day 15, for example, does not automatically indicate that a lunar calendar was in operation. All that is signified is that: a) by sighting, the fifteenth day in the cycle of a moon occurred, or b) the name of the day refers to a civil day and not to a lunar one. Thus, in an inscription about day 23 of civil month two (Phaophi), a lunar-based event could be referred to as "lunar day 6" and the corresponding festival ascribed. In the Kahun and Illahun papyri, as presented by Luft, such occurrences were not rare.

Nevertheless, the existence of a "second lunar calendar" or even the original one operating side-by-side with its civil equivalent is hard to reconcile with the evidence from the New Kingdom festival calendars. Rather, it seems more reasonable to argue that the so-called "festivals of heaven" merely describe those lunar days which coursed through the civil year, without any connection to a specific lunar calendar. The priests simply observed the phases of the moon, from *psḏntjw* up to the twenty-ninth or thirtieth day.[48] This would explain those few indications of lunar determined events in the New Kingdom festival calendars and elsewhere during the same epoch. Earlier, however, as confirmed by Luft's work on the lunar-based *wꜣgj* feast during the late Twelfth Dynasty, such does not appear to be the case.[49]

Yet there remains the nagging supposition that those *ḥbw nw pt* at Medinet Habu are nothing more than the names of the civil days. Consider the introductory remarks in List 7: "The festivals of heaven which occur. An offering to Amun-Re and the *šsm-ḫw* of the King of Upper and Lower Egypt, Woser-ma'at-Re-mery-Amun ... in the course (*ẖrt*) every month."[50] Is it not simpler to assume that these offerings for days 29 and so forth refer to civil days that occurred twelve times per year? By this time, it could be argued, the original significance of these seven had obscured to such a point where these names no longer meant, if always ever, lunar days. Although this is a difficult case to prove, it cannot be refuted categorically, and it remains true that their general nomenclature, "festivals of heaven," always signified events which occurred more than once within a civil year, and nothing more.

A festival stela from Buto dated to the reign of Thutmose III provides some additional evidence with regard to these "heavenly" celebrations.[51] They are not labeled *ḥbw nw pt*. Although recorded separately from the annual feasts which follow — just as at Medinet Habu — the entire group of religious occurrences is called "seasonal

---

48 I first proposed this straightforward solution in "The Lunar System of Festival Calendars: From the New Kingdom Onwards," *BSEG* 19 (1995): 25–40. There was no need for a "second lunar calendar," à la Parker, *Calendars*, pp. 53–54.

49 His work is cited in note 24, above.

50 *KRI* V 130.10–11.

51 Shafia Bedier, "Ein Stiftungsdekret Thutmosis III.," *Bulletin of the Center of Papyrological Studies* 10 (1994): 1–23; add Spalinger, "The Festival Structure of Thutmose III's Buto Stela," *JARCE* 33 (1996): 69–76.

festivals (*tp trw*). We must be extremely careful when analyzing ancient Egyptian nomenclature. Often it is the case that specific terms which ought to occur are left out, and the mundane arithmetic that we would like to have seen is omitted. Yet to allow what appears to be an equal number of day thirties with, say, the same number of day ones, must indicate a civil rather than a lunar setting.

From the wealth of the Kahun and Illahun material Luft was able to reconstruct a detailed account of the yearly religious calendar with the complex interaction of preparations for various lunar and civil feasts. Evidently, the priests at these religious edifices maintained a clear distinction between the two methods of arranging their liturgical calendars and, not surprisingly, the retention of the older lunar celebrations was common. However, no hieroglyphic copy of a festival calendar exists from this region and period to enable us to compare the specific daily accounts with the ideal religious calendar.

All temples appear to have established and were endowed with a regular, indeed mundane, program in which civil dates were recorded along with the corresponding festivals. The Medinet Habu calendar of Ramesses III is the most detailed one from the New Kingdom.[52] There, few lunar-based feasts are logged, except possibly for those related to the phases of the moon: the *ḥbw nw pt*. An open date, so to speak, is normally given for the few annual festivals that were determined by the moon, as in List 3. The Valley Festival was listed separate from the *tp trw* or annual ones. Therein is described the exact timing of this feast: "It is *psḏntjw* which fetches/d it." Only the civil month is given, the second of *šmw*, harvest. The event was determined by the moon but set within a specific civil month in the standard operating calendar. What probably happened was the naked eye sighting of *psḏntjw*, a relatively simple operation. No established and working lunar calendar needs to be hypothesized; the priests did not have to consult their archives to determine which lunar month within a lunar year it was.

The last point is the essential one. It is often argued that side-by-side with the Civil Calendar there persisted the earlier lunar one, which regulated the "moveable" feasts that were still observed in pharaonic times.[53] However, this hypothesis seems unwarranted. The Valley Feast always took place in the second month of *šmw*;[54] the processional feast for Amun was likewise set on the same lunar day but in civil month nine (List 67). There is an apparent contradiction to these two cases. List 63 refers to *psḏntjw* on the first day of month eight — the exact name of the religious event is unclear — and the same phrase, "It is *psḏntjw* which fetches/d it," occurs in List 3. A much later parallel can be seen in Kawa Inscription No. 8 (Macadam's designation),

52 Text conveniently in *KRI* V 115–84.

53 This was alluded to above. The key references are Parker, *Calendars*, pp. 53–54, with Krauss, *Sothis- und Monddaten*, pp. 184–98.

54 For the chronology of the Valley Feast, see Krauss, *Sothis- und Monddaten*, pp. 136–44.

referring to a feast of Amun.[55] Perhaps the most important of these auspicious events can be found in an inscription of Thutmose III, dated to his twenty-fourth regnal year, and this will be addressed in the following discussion.

The interlocking of the original lunar system and the latter solar counterpart is best observed in the exact day upon which certain religious performances took place. Among the most revealing is the so-called "inebriation" feast, *tẖj*. Its position on day 20 of civil month one does not appear to be a result of mere willful choice, as I have earlier indicated. In fact, this feast gave its name to civil month one. On day 19 of the same month, the Egyptians held the important celebration for the moon god Thoth. The purpose of that day reflects the lunar-solar epact, as described above. Even the grandiose Opet feast was not immune to lunar influence.[56] It originally comprised eleven days, a number that should make us wonder whether once more the epact

55 Spalinger, "Chronological Remarks," *BSEG* 22 (1998): 56–57. This corrects a bit of Eleonora Kormyscheva, "Festkalender im Kawa-Tempel," in Rolf Gundlach and Matthias Rochholz, eds., *4. Ägyptologische Tempeltagung: Feste im Tempel* (Wiesbaden, 1998), pp. 77–89.

56 In this context the brief study of William J. Murnane is very useful: "La grande fête d'Opet," *Les Dossiers: histoire et archéologie* 101 (1986): 22–25; see as well Lanny Bell, "Luxor Temple and the Cult of the Royal Ka," *JNES* 44 (1985): 251–94, with his "The New Kingdom 'Divine' Temple: The Example of Luxor," in Schafer, ed., *Temples of Ancient Egypt*, pp. 127–84. Add Erika Meyer-Dietrich, "Die Opetprozession - mehr als nur eine rituelle Verbindung von Karnak mit Luxor: Die Verwendung von Schall zur Erzeugung eines symbolischen Raumes bei der Opetprozession," in I. M. Dolinska and H. Beinlich, eds., *8. Ägyptologische Tempeltagung: Interconnections between Temples* (Wiesbaden: 2010), pp. 123–36. Barry Kemp, in *Ancient Egypt: Anatomy of a Civilization* (London and New York, 1989), pp. 185–97, has presented an excellent overview and analysis of the religious connections inherent in this celebration. I would like to add here some useful comments on little-known yet significant aspects of this feast: Jean-Marie Kruchten, "L'année où la fête d'Opet n'eut pas lieu en Paophi," *JEA* 77 (1991): 181–83, with his *Le grand texte oraculaire de Djéhoutymose intendant du domain d'Amon sous le pontificat de Pinedjem II* (Brussels, 1986), pp. 23–26, 57–71, 246–47, 257–59, 263–64, 310–24, and 344–45. The following studies are connected with data of Ramesses II associated with Luxor: *KRI* II 345–47.4 (background text of regnal year three) and 606–9; Mahmud abd-el Razik, "The Dedicatory Inscription and Building Texts of Ramesses II in Luxor Temple," *JEA* 61 (1975): 125–36; Oleg Berlev, "Bureaucrats," in Sergio Donadoni, ed., *The Egyptians* (Chicago, 1997), pp. 91–92; Amin A. M. A. Amer, "Reflections on the Reign of Ramesses VI," *JEA* 71 (1985): 66–70, with *KRI* VI 340.11–43.10 — either the Valley Festival or that of Opet (The key word *n*[ᶜ …] is in *KRI* II 342.1.); Masashi Fukaya, "Oracular Sessions and the Installations of Priests and Officials at the Opet Festival," *Orient* 47 (2012): 191–211, and *Socio-Religious Functions of Three Theban Festivals in the New Kingdom: The Festivals of Opet, the Valley, and the New Year* (Oxford, PhD Thesis, 2014).

of the sun and moon was in consideration.[57] From the accounts of Thutmose III it was set on day 15, very conveniently the mid-point of the second civil month. And yet we find the commencement of this event displaced in later times. By the reign of Ramesses II, as List 29 of the Medinet Habu festival calendar confirms, Opet had expanded to twenty-four days, from the nineteenth of the second civil month until the twelfth of the third. (See the use of the integer 19 once more; it is the "other half" or antithesis of 11.)

Corresponding to the great development of the Opet feast was that of the Valley.[58] We can trace this celebration back to the Middle Kingdom, when it took place in the first month of the harvest season. A graffito of a local Theban priest states: "Giving praise to Amun, kissing the earth to the lord of the gods in his feasts of the first of *šmw* when he arises; the day of the navigation to the Valley of Nebhepetre."[59] (A second reference in the reign of Sesostris III may allude to this festival, but there is no date.)[60] However, as Medinet Habu indicates, in the New Kingdom the festival of the Valley now took place in the second month of harvest. Can we maintain that this portion of the inscription is contemporary with Ramesses III? If so, then additional Twentieth Dynasty material dating this event occurring in the same month makes perfect sense, for the lunar event now occurred later. Yet if the sighting of lunar day 1 was all that important, is it not easier to surmise that the celebration was lunar-based but subsequently re-set in the second month of *šmw*?[61]

## C. FEASTS IN CALENDARS

For the Old Kingdom we have little information concerning these lunar-civil equivalences. *W3gj* is perhaps the best case that can be mustered, and even here scholarly disputes have arisen. Wolfgang Helck, who presented a remarkable, if speculative,

57 *Urk.* IV 742.1 with 824.10; see the cautionary remarks of Gardiner, "Thutmosis III Returns Thanks to Amun," p. 22 with note 3.

58 In general, Siegfried Schott, *Das schöne Fest vom Wüstentale: Festbräuche einer Totenstadt.* Akademie der Wissenschaften und der Literatur, Abhandlungen der Geistes- und Sozialwissenschaftlichen Klasse 1952.11 (Wiesbaden, 1953); add Silvia Wiebach, "Die Begegnung von Lebenden und Toten im Rahmen des thebanischen Talfestes," *SAK* 13 (1986): 263–91.

59 Herbert Winlock, *The Rise and Fall of the Middle Kingdom at Thebes* (New York, 1947), pl. 40, no. 1. El-Sabban refers to this text on page 10 of his *Temple Festival Calendars*; see Krauss *Sothis- und Monddaten*, pp. 136–44, but no final calendrical certainty is provided by the datum.

60 Edouard Naville, *The XIth Dynasty Temple of Deir el-Bahari* I (London, 1907), pl. XXIV.

61 Depuydt addresses this situation in great detail in the study, *Civil Calendar and Lunar Calendar in Ancient Egypt.* Oddly, he fails to take into account the research of Wells,

reconstruction of Niuserre's sun temple calendar, had faced the same problem of lunar dating.[62] It is significant that some of the feasts can be seen in the Middle Kingdom via the Kahun and Illahun material—*šsp jtrw* and *sꜣḏ*, for example, but not later.[63] Of course, the former of the two was well-known from Old Kingdom times.[64] Other celebrations reconstructed by Helck included that of Sokar, *tpj rnpt*[65] *wꜣḥ ꜥḫ*, the Min festival, "clothing," *mnḫt*, as well as a series of better-known events: *wꜣgj*, Thoth, *rkḥ*, and *tḫj* (the last oddly not fully understood by Helck).[66] Despite the fragmentary nature of the material, as well as the bold and daring reconstruction by Helck, it seems reasonably clear that Niuserre's sun temple calendar reflected the age in which it was composed and operated rather well. See in particular the paralleling of some of the events with those on the Old Kingdom private festival lists. But as for specific lunar-based religious observations, we cannot proceed deeper. This is particularly unfortunate as there is present the Min feast, an event well-known to be determined by the moon.

The private feasts lists dateable to the Old Kingdom imply an older system of religious observance in which the Thoth feast preceded that of *wꜣgj*. It seems reasonable

---

who is never cited in the bibliography. The sun's rebirth in the winter solstice is a crucial matter.

62 Wolfgang Helck, "Die 'Weihinschrift' aus dem Taltempel der Sonnenheiligtums des Königs Neuserre bei Abu Gurob," *SAK* 5 (1977): 47–77. The entire reconstruction is, however, dubious. At the minimum, the feasts of *wꜣgj* and Thoth took place in the Civil Calendar on days 18 and 19 of the first month, not on days 19 and 20 as Helck maintained. Nigel C. Strudwick's translation, *Texts from the Pyramid Age* (Atlanta, 2005), pp. 86–91, further confuses the entire matter by following Helck (and slipping an integer to boot: read 100,800 for 108,000 on page 89). In column 7 on plate II, Helck restores "columns," but the temple is known to have had only two, as Prof. Miroslav Verner informs me.

63 See Luft, *Die chronologische Fixierung des ägyptischen Mittleren Reiches*, pp. 183–88, and his "Das *sꜣḏ*-Fest im Alten Reich," in Daniela Mendel und Ulrike Claudi, eds., *Ägypten im afro-orientalischen Kontext: Aufsätze zur Archäologie, Geschichte und Sprache eines unbegrenzten Raumes: Gedenkschrift Peter Behrens* (Cologne, 1991), pp. 237–44.

64 In particular, see the Old Kingdom feast lists: Spalinger, *The Private Feast Lists of Ancient Egypt*, Chapters 1 and 6.

65 Helck partly misunderstood this event. Apparently, so did Willems, *The Coffin of Heqata (Cairo JdE 36418)*, p. 356, note 2195. *Tpj rnpt* is a new year's day, but not the all-important New Year's Day of I *ꜣḫt* 1. In the Civil Calendar the latter is designated by *wp rnpt*. The issue of *tpj rnpt* will be covered later.

66 These are all covered by Luft in his *Die chronologische Fixierung des ägyptischen Mittleren Reiches*. One of the most important for us is, of course, *mnḫt*, as it was the name of an ancient Egyptian lunar month (III) and later transferred "back" to civil month II; see Luft's analysis on pages 160–63.

to conclude that the latter event is none other than the original, ancient, moveable lunar-based *wꜣgj*. The civilly determined *wꜣgj* occurred one day earlier than the fixed Thoth feast, the latter set permanently on day nineteen of civil month one. Thus the private feast lists reflect an era in which the civil *wꜣgj* had not gained precedence. Notwithstanding this conclusion, those non-royal lists seem to work on a schedule that courses through a year, though what type of year was *originally* intended remains unclear. Nonetheless, if we pay attention to the situation of Thoth preceding *wꜣgj*, a good argument could be made for the original lunar calendar.

Moreover, the separation of New Year's Day, *wp rnpt*, from a second, *tpj rnpt*, is hard to resolve unless we regard the latter as the very old designation for the festival of Sothis, later described as the "going forth of Sothis," *prt Spdt*.[67] This is one reason why Helck placed *tpj rnpt* on the first days of the second season, I *prt* 1. He had come to the conclusion that *tpj rnpt* signified the *renovatio temporum*, the rebirth of a cycle, and also felt — correctly I would argue — that the same day was the ideal one for the commencement of the pharaoh's heb sed. Further support for this analysis is forthcoming from indirect means.[68] It is not until the Twelfth Dynasty that we hear of the festival *prt Spdt*, whereas the oft spoken "second New Year," *tpj rnpt*, is very common in the Old as well as Middle Kingdoms. Indeed, it is striking that *prt Spdt* is unmentioned either in the private feast lists of the Old Kingdom or in one partly extant festival calendar dating from the same time frame, that of Niuserre.

67 Parker, *Calendars*, pp. 61–62, attempted a solution, but this is no longer followed. Helck wished to make it coincide with the Nehebkau celebration on the first day of the second season, I *prt* 1: "Die 'Weihinschrift' aus dem Taltempel des Sonnenheiligtums des Königs Neuserre bei Abu Gurob," pp. 51 (9,d) and 60. See my comments in "A Remark on Renewal," *SAK* 17 (1990): 289–94, and "Calendars: Real and Ideal," in Betsy Bryan and David Lorton, eds., *Essays in Egyptology in Honor of Hans Goedicke* (San Antonio, 1994), pp. 297–308. For the connection with the heb sed, see the summary of Jocelyn Gohary, *Akhenaton's Sed-Festival at Karnak* (London and New York, 1992), p. 4, and later Susanne Voß, *Untersuchungen zu den Sonnenheiligtümern der 5. Dynastie: Bedeutung und Funktion eines singulären Tempeltyps im Alten Reich* (Hamburg, Ph.D. Dissertation, 2004), pp. 124–34. A good argument with regard to the assumed accession of Akhenaton can be made for I *prt* 1. In general, see William Murnane, "On the Accession Date of Akhenaten," in Johnson and Wente, eds., *Studies in Honor of George R. Hughes*, pp. 163–67; and with Charles Van Siclen III, *The Boundary Stelae of Akhenaton* (London and New York, 1993), pp. 149–50. This key date, however, may indicate the new heb sed festival of the Aten. Additional commentary may be found in Chapter 3, note 65.

68 I analyzed the situation of *tpj rnpt* further in *The Private Feast Lists of Ancient Egypt*, passim, and especially chapter 2.

Independently of the lunar evidence from the Old Kingdom, there is an interesting one from Coptos and dating to the Eleventh Dynasty.[69] The stela containing the key reference reads: "Regnal year one, fourth month of Inundation, day 25, lunar day 15. (Note once more the importance of the mid-lunar month.) The method of writing the date is somewhat archaic, but the equation is clear. The next instances are to be found in the later Twelfth Dynasty papyri from Illahun and Kahun, as well as some private inscriptions of the Middle Kingdom. Very rarely do we encounter contemporary lunar references in the royal material. The famous one of Thutmose III's twenty-third regnal year, where the king describes his battle at Megiddo, placed the first lunar day (once more!) to day 21 in the ninth civil month.[70]

One major source of documentation is the inscriptional material in the festival calendars. They are an invaluable source for many subfields of Egyptology: economics, religion, and festivals, besides calendrics. In the earliest known calendar, that of Niuserre, we find the more primitive format, without boxes and a grid pattern, typical of the New Kingdom.[71] Nevertheless, all of the major components of the later festival calendars can be discerned. In Niuserre's text the day of New Year's (*wp rnpt)* is explicitly different from *tpj rnpt*, a further point that was in favor of Helck's reconstruction. In addition, the separation of the annual offerings from those that occurred with more frequency, which we find in later festival calendars, must have been intentional. Yet the inclusion of specific lunar organized feasts is striking, as

69 This Coptos reference to Sothis is referred to later in Chapter 3, note 89. It will be found in Hans Goedicke, "A Cult Inventory of the Eighth Dynasty from Coptos (Cairo JE 43290)," *MDAIK* 50 (1994): 71–84; Fischer, "Notes on Some Texts of the Old Kingdom and Later," in Peter der Manuelian, ed., *Studies in Honor of William Kelly Simpson* I (Boston, 1996), pp. 267–70; and Krauss, "Ein Versuch zur Chronologie des späten Alten Reiches im Anschluss an die Monddaten im Neferirkare-Archiv." Add the Gebel Tjauti Rock Inscription 11 for yet another heliacal rising of Sothis observation: John Coleman Darnell, *Theban Desert Road Survey in the Egyptian Western Desert. Volume 1: Gebel Tjauti Rock Inscriptions 1–45 and Wadi el-Hôl Rock Inscriptions 1–45* (Chicago, 2002), pp. 49–51. The day is not included in the date and the king associated with the given regnal year eleven is also not named. Darnell places the text in the Seventeenth Dynasty. We must remember that the official determination of the heliacal rising of Sothis (at Heliopolis and by the priesthood) determined the official Egyptian chronology. One more useful reference to the heliacal rising of Sothis may be found in Janine Bourriau, "Three Monuments from Memphis in the Fitzwilliam Museum," *JEA* 62 (1982): 53 — awaiting for the *hrw prt Spdt*, with Luft, "Illahunstudien IV: Zur chronologischen Verwertbarkeit des Sothisdatums," *SAK* 16 (1989): 230 and note 45.

70 *Urk.* IV 657.2.

71 Helck, "Die 'Weihinschrift' aus dem Taltempel des Sonnenheiligtums des Königs Neuserre bei Abu Gurob," and el-Sabban, *Temple Festival Calendars of Ancient Egypt*, pp. 2–8.

we have already noticed with the Procession of Min (*prt Mn*)[72] and the celebration of *sꜣḏ*, neither of which appears to have been permanently regularized in the Civil Calendar. Both of these events occur in the Old Kingdom private feast lists, thus indicating that the latter and Niuserre's calendar are contemporaneous.

The private feast lists changed over time. The Middle Kingdom arrangement is more haphazard than the Old Kingdom one, and by the New Kingdom the Egyptians drew up their lists in a totally different fashion.[73] Yet one important change can be witnessed from the end of the Sixth Dynasty. The chronological arrangement now followed the civilly based *wꜣgj*, because the latter preceded rather than followed Thoth, and this fits their known civil dates in the first month: 18 for *wꜣgj* and 19 for Thoth. Evidently, this small change further indicates the ever-increasing importance of the second Egyptian annual calendar.

It is from the New Kingdom, however, that far more information is forthcoming. An inscription in Nubia, dated to Thutmose III's second regnal year, is of particular importance.[74] This text was drawn up to reinvigorate or reinforce an earlier decree of Sesostris III of the Twelfth Dynasty. The decree is not organized according to the format of this era. Calendrical references are more narrative than arithmetically placed. A better example of the New Kingdom approach can be seen in Thutmose III's feast list from Karnak.[75] It is short and compartmentalized. Horizontal rectangles contain the name of each feast. To the right will be found the number of days each event was celebrated, followed by more boxes in which various food totals were written down. In the list, the Nehebkau feast, celebrated on day 1 of the second season, is overtly separated from the religious occurrences that precede. Moreover, the final rite at the temple comprises two moon days, the first and the sixth. Here, the foods are divided evenly by twelve, suggesting again that the presumed lunar days were either just civil ones — days 1 and 6 of the twelve civil months in this case — or else simply the lunar days irrespective of any organized and independent lunar calendar. There is a crucial problem to this reference. Days 1 and 6 are supposed to have been celebrated at least eighteen if not twenty-eight times in the (civil) year, either being a definite impossibility. One would have expected twenty-four, or possibly twenty-

72 Parker, *Calendars*, pp. 36, 39–40, 47–48 (now dated), and Krauss, *Sothis- und Monddaten*, pp. 142–43.

73 *The Private Feast Lists*, chapters 2–3.

74 See now Ricardo Caminos, *Semna-Kumma I: The Temple of Semna* (London, 1998), pp. 41–48, with pls. 23, 25–26; older edition in *Urk.* IV 193.13–196.9. Add Parker, *Calendars*, pp. 33–34, and Spalinger, *The Private Feast Lists*, pp. 6–8.

75 *Urk.* IV 177, and el-Sabban, *Temple Festival Calendars*, p. 19. Some corrections to Sethe's edition have to be made which are not recognized by el-Sabban: Spalinger, "The Calendrical Importance of the Tombos Stela," *SAK* 22 (1995): 279–81, and pl. 12; see already Parker, *BiOr* 9 (1952): 101, in the review of Schott, *Altägyptische Festdaten*.

six, if lunar-based and operating with thirteen lunar months. Is there an error here with 10 + 4 and [10] + 4 incorrectly written for 10 + 2 and [10] + 2?

At Thutmose III's festival temples there was an additional calendar, which has already been discussed because of its importance for the Civil Calendar.[76] The organization and structure parallel his Karnak list, even though only the annual events (the *tp trw*) are covered. In contrast, his Elephantine Decree, while fragmentary, is very significant for the mention of the heliacal rising of Sothis, this time placed on day 28 of the eleventh month.[77]

It is also essential to Krauss' hypothesis of the originals of the Civil Calendar. But are all references to this event contemporary? At Medinet Habu, to take a care in point, the festival calendar merely indicates that *prt Spdt* is to be celebrated in the first month of the year; no day is given.[78] Should we understand the reference as a global and self-understood one? That is to say, is only I *ꜣḫt* understood? Thutmose III's Buto calendar contains a striking case of a non sequitur. The day is indicated "at its times."[79] A similar notation appears in the fragmentary calendar of Amunhotep I from Karnak. In both of these cases we can figure out that the exact date was probably unknown. After all, it moved slowly in the Civil Calendar, and the exact timing depended upon the latitude in which the star was sighted. Then there is always the possibility that a reference to the heliacal rising of Sothis may reflect an earlier time if that festival calendar was a reworking of a prior decree; see Medinet Habu, for example.

From the end of the Eighteenth Dynasty fragments of offerings lists of Amunhotep IV/Akhenaton remain, even though not completely published, to reconstruct a similar outlook.[80] Whatever economic data the new decrees provide, Akhenaton's sun

76 *Urk.* IV 1272.4–1273; Gardiner, "Thutmosis III Returns Thanks to Amun," pp. 20–23; and el-Sabban, *Temple Festival Calendars*, pp. 29–30.

77 The fragment containing this event is nicely photographed in Jean-Paul Boulanger and Geneviève Rensio, *Naissance de l'écriture: cunéiformes et hiéroglyphes: [exposition], Galeries nationales du Grand Palais, 7 mai-9 août 1982* (Paris, 1982), p. 265 (Christine Ziegler); see *Urk.* IV 823–9.4 with 827.8–9 for *prt Spdt*, and el-Sabban, *Temple Festival Calendars*, pp. 31–37.

78 *KRI* V 144.6; see Parker, *Calendars*, p. 40, and Spalinger, "Sothis and 'Official' Calendar Texts," *Varia Aegyptia* 10 (1995 = 1997): 175–83 for the following discussion of "peculiar" *prt Spdt* dates. Add Caminos, *Semna-Kumma* I, p. 46 and note 3, a useful commentary on this matter.

79 Bedier, "Ein Stiftungsdekret Thutmosis' III.," p. 17; Spalinger, "The Festival Structure of Thutmose III's Buto Stela."

80 Ramadan Saad and Lise Manniche, "A Unique Offering List of Amenophis IV Recently Found at Karnak," *JEA* 57 (1971): 70–72; Helck, "Zur Opferliste Amenophis' IV. (*JEA* 57, 70ff.)," *JEA* 59 (1973): 95–99; Claude Traunecker, "Données nouvelles sur le début du règne d'Aménophis IV et son œuvre à Karnak," *JSSEA* 14 (1984): 60–69; and "Amen-

cult deleted the age-old calendrical events. The dates must have been, by nature, very different from the older ones, whether lunar or civilly determined. Not surprisingly, in the Ramesside period the earlier formats were reinstated. Ramesses II's calendar from Abydos is illustrative in this context even though it reflects the aspects of the Osirian rituals peculiar to this key site.[81] An endowment stela from Thebes, also dated to this king's reign, is useful for economic analysis; the same is true of a festal stela of Ramesses IV.[82] It is the great calendar at Medinet Habu that is our primary source of information on the calendrical timing of feasts — both civil and lunar — and the internal order of the Egyptian year.[83] The arrangement is as one might suspect: the commencement of the civil year is signaled by *wp rnpt,* and the lists of religious observances continue up to the last feast celebrated (List 67: in the first month of *šmw*: cut here owing to space limitations). The box-like arrangement is conspicuous, as is the division between annual and repetitive celebrations. Separate from the main body are, significantly, the lunar-based Valley Feast, the regular morning offerings, and new feasts established by Ramesses III. There are two secular events as well, a feast of victory over an unknown foe (List 5) and a commemoration of the repulsion of the Meshwesh in Ramesses III's eleventh regnal year (Lists 53–55: new addition). There are many "topographical" errors in this lengthy text, which should make us wary of accepting any date without skepticism. To take a case in point, the two lists 24–25, referring to the *wꜣgj* feast, stipulate days 18 and 19 of the first month; actually, we have to subtract one day from both. The lunar references are, as we have seen, few and far between, a situation that became even more common in the post-Empire era.

The final group of calendars come from the Graeco-Roman Period. Fortunately, all have been subject to minute analysis by Alfred Grimm.[84] Once more the entire

---

hotep IV, percepteur royal du Disque," in Thierry-Louis Bergerot, ed., *Akhénaton et l'époque amarnienne* (Paris, 2005), pp. 145–82. Note that on page 168 of the second work, Traunecker notes fragments of Akhenaton from Karnak referring to "chariot of the feast of Choiak," "chariots of the New Year," and "List of the New Year."

81 *KRI* II 513–16; el-Sabban, *Temple Festival Calendars*, pp. 38–52; and Spalinger, "The Grain System of Dynasty 18," *SAK* 14 (1987): 305–6, for the grain totals and arithmetical calculations that el-Sabban misunderstood.

82 For the first: *KRI* II 710; Nora Scott, "Recent Additions to the Egyptian Collection," *BMMA* 15 (NS) (1956): 84 (fig. 11); and Michel Gitton, *Les divines épouses de la 18e dynastie* (Paris, 1984), pp. 33–34. For the second: *KRI* VI 7–9; Helck, "Ramessidische Inschriften aus Karnak," *ZÄS* 82 (1956): 98–109; and el-Sabban, *Temple Festival Calendars*, pp. 151–53.

83 See the references provided in note 40, above.

84 Alfred Grimm, *Die altägyptischen Festkalender in den Tempeln der griechisch-römischen Epoche* (Wiesbaden, 1994). Bernard Arquier and Nandine Guilhou, "Cette obscure clarté qui tombe des étoiles. Les fêtes de fondation et de dédicace du temple d'Edfou," in Amanda-Alice Maravelia, ed., *En quête de la lumière. In Quest of Light:*

system has been organized civilly. Now, however, the economic aspects of the various rituals are not included. The calendars from Edfu, Dendera, Kom Ombo, and Esna attest to processions or appearances of the local gods. To be sure, a visit of the deities of Edfu and Dendera is the high point of the religious year. As for the lunar feasts, there is no separation between the annual events and those repeated more than once per year, as we have noted earlier by viewing the contrast between the *tp trw* feasts and the *ḥbw nw pt*. Hence, there are no references to "festivals of heaven," the New Kingdom day feasts. Some lunar regulated celebrations are listed, but they are few.

The festival calendars provide the major source of information about ancient Egyptian religious calendrics. Other texts, of a considerably more mundane nature, shed tantalizing clues. There remains, however, one Middle Kingdom source which supplements our conclusion; namely, the Kahun papyrus.[85] This document, in conjunction with the plethora of fragmentary papyri from Illahun, provides a deeper understanding of the regularity of the Egyptians' method of calendrical reckoning.

There is a major Kahun fragment, written in hieratic of course, which presents the same yearly organization that is to be found in the hieroglyphic festival calendars. But the Kahun list goes into more detail than any standard temple calendar would, although it is lacking the economic endowment of foods. In particular, various individuals associated with festivals are mentioned; e.g., singers, phylae members,[86] and

---

*Mélanges in Honorem Ashraf A. Sadek* (Oxford, 2009), pp. 3–18 is an interesting discussion of the lunar connection of foundation dates from the temple of Edfu. It needed more calendrical know-how.

85 The following is an all-too brief summary of Luft's magnificent work *Die chronologische Fixierung des ägyptischen Mittleren Reiches.*

86 Ann Macy Roth, *Egyptian Phyles in the Old Kingdom: The Evolution of a System of Social Organization* (Chicago, 1991), pp. 3–7 covers the development of the phylae system. Cf. the review of Paule Posener-Kriéger in *CdE* 71 (1996): 73–79, and pages 74–76 in particular. For the Middle Kingdom add: Katalin Anna Kóthay, "Phyles of Stone-Workers in the Phylae System of the Middle Kingdom," *ZÄS* 134 (2007): 138–50. The Middle Kingdom did not revert to the earlier lunar-based system predating the Third Dynasty. The rotation of phylae remained set within the Civil Calendar, a fact that is so clearly evidenced in the famous contract texts of Hekanakht. Roth has unfortunately followed Borchardt's analysis. See now "Praise God and Pay the Priests," *ÄuL* 7 (1998): 43–57; add Jac. J. Janssen, "Rations with Riddles," *GM* 124 (1991): 91–97. The division of the civil year by five phylae was no problem to the Egyptians. It allowed for a 5-year cycle and solved the arrangement for the epagomenal days. The later switch to four phylae undoubtedly was a "cleaner" way of arranging the priestly rotation. It is clear that there were originally eleven months in the predynastic lunar year because the two months called "Great Heat" and "Small Heat" earlier were one; cf. Parker, *Calendars*, p. 7, note 100. Hutton Webster, *Rest Days: A Study in Early Law and Morality* (New York, 1916), p. 177, note 1, raises this point (a year comprising ten months) in another context. But see Jörg Rüpke, *Kalender und Öffentlichkeit. Die Geschichte der Repräsentation und*

others. The specific dates included prove that the papyrus fragment is a contemporary account of the working religious year.[87]

It is from such accounts, especially the small fragments coming from Illahun, that we can reconstruct the workings of Egyptian calendrics. At Illahun, quite a number of papyri deal with these matters in the most mundane of circumstances, ranging from the payroll for the phylae priests to explicit timetables for civil and lunar events. There is an important reference to the heliacal rising of Sothis, this time from a contemporary and non-hieroglyphic source.

New Kingdom ostraca found in the workmen's village of Deir el Medineh contain references to various Egyptian festivals, data which have enabled scholars to piece together details of Egyptian calendrics.[88] Such "practical" or "workaday" documents confirm the rather simplistic system of calendrical notation, the dominance of the civil year, and the rarity of lunar-based events. Moreover, one can follow the calculations as they were made, even though many are quite minute in operation.

The emphasis given in this discussion to religious matters — particularly with respect to religious festivals — indicates the limits of our sources. In fact, it is the case that supplementary material is to a large extent non-secular. The famous series of water clocks, for example, all organized to a 360 day civil year, was used for timing religious performances and duties. The truncated year, in which the civil one lost the five epagomenal days, was adopted for simplicity and regularity.[89]

---

*religiösen Qualifikation von Zeit in Rom* (Berlin and New York, 1995), pp. 192–202; add Robert Hannah, *Greek and Roman Calendars. Constructions of Time in the Classical World* (London, 2005), p. 99.

87 It also provides welcome data concerning the *jhhj* feast; see Grimm, "Zur kalendarischen Fixierung des *jhhy*-(Freuden-)Festes nach dem Festkalender des Königs Amenophis I. aus Karnak," *GM* 143 (1994): 73–76. Its lunar associations have yet to be defined precisely.

88 Helck, "Feiertage und Arbeitstage in der Ramessidenzeit," *JESHO* 7 (1964): 136–66; R. van Walsem, "Month-Names and Feasts at Deir el-Medîna," in R. J. Demarée and Jac. J. Janssen, eds., *Gleanings from Deir el Medîna* (Leiden, 1982), pp. 215–44; and Heidi Wikgren, "The Festival Calendar at Deir el-Medina," in Kathryn Piquette and Serena Love, eds., *Current Research in Egyptology 2003* (Oxford, 2005), 179–200. The last is incomplete insofar as she does not cover the lunar events as perceived and dealt with by the Egyptians. The same may be said with respect to Heidi Jauhiainen, *"Do Not Celebrate Your Feast Without Your Neighbours": A Study of References to Feasts and Festivals in Non-Literary Documents from Ramesside Period Deir el-Medina* (Helskinki, Ph.D. Thesis, 2009). For additional commentary, see Fredrik Hagen and Henriette Koefoed, "Private Feasts at Deir el-Medina: Aspects of Eating and Drinking in an Ancient Egyptian Village," *Archaeological Review from Cambridge* 20 (2005): 6–31.

89 The 360 day truncated year is often misrepresented as something astronomical. It was not. The simpler division that could be performed with 360 instead of 365, coupled

## D. THE EGYPTIAN HOURS AND TIMEKEEPING[90]

The Egyptians chose a system of measuring what we call hours by addressing the situation of the night. But how could they divide the time when the sun, the best marker, was absent? One could not tell midnight like midday, even approximately, when there were no shadows. Therefore the Egyptians, as other people, looked at the stars. Owing to their concentration upon the eastern horizon, so overtly proved by their emphasis on the heliacal rising of Sothis, the rising of stars in the east formed the basis of the nocturnal hour reckoning. True, the moon could be observed, but in essence its motion (both in longitude and in latitude) was useful for only larger intervals: days and months. I suspect that the Egyptians may not have paid strict attention to morning or evening twilight in predynastic times, but by historical times these two events were recognized, and the beginning of the day seems to have commenced with the latter phenomenon instead of the actual rise of the sun over the eastern horizon.[91]

---

with the twelve-month civil year led to its rapid adoption for accounts, payments, and the like. The subject is worth further research, but for the moment, George A. Reisner, "The Tomb of Hepzefa, Nomarch of Siût," *JEA* 5 (1918): 84 is the *locus classicus* discussion which defines a "temple day" (= 1/360 of a year). On the Manshiyet es-Sadr Stela of Ramesses II (*KRI* II 361.5–6), the period of time in which the stone workers performed their task covered "one year and three days." However, the latter excluded the epagomenal days: Jürgen von Beckerath, "Das Thronbesteigungsdatum Ramses II," *ZÄS* 81 (1956): 1. For the later Ramesside Period and the payments during the epagomenal days, see now Janssen, "Rations with Riddles," *GM* 124 (1991): 91–97, and "Rations with Riddles II," *GM* 128 (1992): 81–86. Lis Brack-Bernsen, "The 360-Day Year in Mesopotamia," in Steele, ed., *Calendars and Years*, pp. 83–100, is useful to consult for comparative data.

90 The approach taken here is historical and scientific. Thus it diverges considerably from the religious-philosophical perspectives of Jan Assmann, *Steinzeit und Sternzeit: Altägyptische Zeitkonzepte* (Munich, 2011), and his earlier study, *Zeit und Ewigkeit: Ein Beitrag zur Geschichte der Ewigkeit* (Heidelberg, 1975).

91 This has been the contention of Krauss and Wells but not that of Luft and Leitz. I have provided additional data in support of the morning twilight epoch in "Nut and the Egyptologists," *SAK* 41 (2012): 353–77. Part of that analysis is presented here in abbreviated format. One might add Richard Jasnow, "On Sothis and the So-Called Clepsydra (?) in the Book of Thoth (Bo2, 3/16–4/11)," in Hermann Knuf, Christian Leitz, and Daniel von Recklinghausen (eds.), *Honi soit qui mal y pense: Studien zum pharaonischen, griechisch-römischen und spätantiken Ägypten zu Ehren von Heinz-Josef Thissen* (Leuven, Paris, and Walpole MA, 2010), pp. 237–43. Some of the commentary is pertinent to our discussion. He also confronted, but did not understand, the difference between *tpy rnpt* and *wp rnpt*. An old study can still be referred to: E. Vassy, "La lumière du ciel nocturne," *BIE* 33 (1950–1951): 203–6. But I can add Jaan Puhvel, "Eventide in

The earliest system of telling time in the night appears to have been derived from a religious influence. Although the earlier useful source material is dated to the Middle Kingdom, it is unmistakable when we read these texts — the so-called decanal system of stars on coffins — that a complete system of stellar reckoning for the night hours had already evolved. The so-called "diagonal star clocks" must have been invented during an era of political and social stability and not one of war and anarchy. Indeed, the most recent analysis of this material places the diagonal star clocks to the Old Kingdom.[92] In addition, the system was well-suited for the institution of the priesthood. Certain rites would have been performed at various times in the night and not merely at sunset or sunrise, morning or evening twilight. How could they be regularized?

Coffins, now dated to the Middle Kingdom, contain a series of diagonally oriented texts in which various star combinations, plus some individual stars, determined the major divisions (hours) of the night. None of these hours were of equal time in our sense. Indeed, the segments — twelve in number — were not of the same length throughout the year. Nevertheless, this method of timekeeping was adequate for the needs of the priestly class.

It was the risings of these chosen star groups, called "decans" in Egyptological literature, that determined the specific intervals of the night. Scholars still have not worked out all the details even if it seems, following Quack,[93] that their invention occurred in the middle of the third millennium BC. The majority opinion is that the rising of a decan marked the end of an hour, and the orientation to the east — as emphasized more than once above — should not be surprising to us. After all, the crucial timing of Sothis is an eastern event, and the moon's last crescent disappears in the east. Fortunately, supplementary information from the cenotaph of Seti I at Abydos as well as the "Nut Book," reedited by Alexandra von Lieven, help us to reconstruct the essentials of the decanal system and its presuppositions.[94] Moreover, with the recent reevaluation of the key diagonal sector in these lists that concerns the epagomenal days, it can now be argued strongly that the decanal system of observation is, indeed, older than the First Intermediate Period.

---

Hatti-land," in Brigitte L. M. Bauer and Georges-Jean Pinault, eds, *Language in Time and Space. A Festschrift for Werner Winter on the Occasion of his 80th Birthday* (Berlin and New York, 2003), pp. 347–49; and T. L. Markey, "Morning, Evening, and the Twilight Between," in Susan Nacev Skomal and Edgar C. Polomé, eds., *Proto-Indo-European. The Archaeology of a Linguistic Problem: Studies in Honor of Marija Gimbutas* (Washington D.C., 1987), pp. 299–321.

92 This has been argued by Prof. Joachim Quack whose research on this important matter will appear in the future. See notes 12–13, above.

93 *Untersuchungen zu den ägyptischen Dekanen und ihrer Rezeption in der griechisch-römischen Welt.*

94 *Grundriss des Laufes der Sterne: Das sogenannte Nutbuch.*

Otto Neugebauer and Richard Parker remarked that in the Seti cenotaph, Re remains outside of Nut until the third night hour. The crucial Seti I texts indicate the following:[95]

| | |
|---|---|
| First Hour: | Re "enters"; *wẖꜣ* (or *ẖꜣwy*)[96] |
| Second Hour: | Re "goes to rest"; *bkꜣt* ("early night")[97] |

On the accompanying shadow clock in the cenotaph, a more specific means of locating the night hours is given.[98] This must refer, as others have seen, to the figure of Nut.

| | |
|---|---|
| First Hour: | "hand" (*ḏrt*); *wẖꜣ* or *ẖꜣwy* |
| Second Hour: | "lip" (*spt*); *wẖꜣ* |
| Third Hour: | "tooth" (*nḥdt*); *wẖꜣ* |
| Fourth Hour: | "throat" (*ḥtt*); *wẖꜣ* |

And so forth.

On the Osireion Star Clock this arrangement is followed:[99]

| | |
|---|---|
| First hour of the evening: | *wẖꜣ/ẖꜣwy* |
| Second hour in early night: | *bkꜣt* |
| Third hour in early night: | *bkꜣt* |
| Fourth hour in early night: | *bkꜣt* |
| Middle night (hour five): | *wšꜣw* |
| Beautiful hour in middle night(hour six): | *wšꜣw* |

95 Neugebauer and Parker, *Egyptian Astronomical Texts I*, p. 82. The main text and scene is to be found in Henri Frankfort, *The Cenotaph of Seti I at Abydos* II (London: Egypt Exploration Society; 1933), pl. LXXXII.

96 A brief remark on the writing is to be found in von Lieven, *Grundriss des Laufes der Sterne*, p. 73, note 353.

97 Ibid., p. 151 (under § 63).

98 Concerning the ancient Egyptian shadow clocks, see most recently Mengoli, *Astronomica Egizia*, pp. 326–50. Add the earlier study of Willibald Henne, "Bemerkungen zur altägyptischen Streiflicht-Uhr," *ZÄS* 118 (1991): 21–31.

99 Neugebauer and Parker, *Egyptian Astronomical Texts I*, pp. 32–33; Margaret A. Murray, *The Osireion at Abydos* (London, 1904), pl. 21 and pl. XII. Von Lieven has discussed the decorative program of this important edifice in "Bemerkungen zum Dekorationsprogramm des Osireion in Abydos," in Haring and Klug, eds., *6. Ägyptologische Tempeltagung: Funktion und Gebrauch altägyptischer Tempelräume, Leiden 4.–7. September 2002* (Wiesbaden, 2007), pp. 167–86.

Therefore, differing designations for the evening hours apparently existed, a point first emphasized by Parker and Neugebauer but also covered by Erik Hornung in 1956.[100] The shadow clock text on the west side of the roof of the sarcophagus chamber in Seti's cenotaph makes reference only to three divisions in the evening or night: *ẖ3wy*, *wš3w*, and *3ẖw* or *wbnw*—"early night," "deep night," and "sunshine" or "(sun) rise." Neugebauer and Parker also felt that the cenotaph divisions of night were symmetrical and could form two halves, with the second parallel to the first.[101] Thus *wš3w* referred to the four-hour separation of "deep night" between *wẖ3/ẖ3wy* (one hour: 1) and *bk3t* (three hours: 2–4). Finally, they felt that a logical division after "middle night" *(wš3w)*, consisting of three hours (9–11), followed by a final one (12), made sense.

In her detailed commentary, von Lieven makes several arguments in order to support an Old Kingdom original.[102] Neugebauer and Parker remarked that in one of the two versions Re "goes directly into the mouth of Nut in the first hour of the evening and another in which the setting took place near Nut's hands and the actual entrance into the Duat through her mouth is later."[103] In addition, they also felt that there were two versions of the rising of the sun. One account has Re leaving the vagina of Nut at the conclusion of night hour nine, whereas in a second version, identified in the "Book of Him who is in the Duat," Re goes forth in hour twelve.

By bringing into light the evidence from a shadow clock, also located in the Seti cenotaph, Neugebauer and Parker were able to demonstrate that one account of the night hours implied the 3–6–3 division.[104] The sun stayed outside the mouth of Nut for the first two hours of the night; he then spent up to hour nine in darkness. At the point Re is commanded to begin to withdraw from the Duat in the hour *sḥtp.n.s*, the exact time is given: the ninth hour (P. Carlsberg I, II 9–11). Then he withdraws through the goddess' vulva in hour ten, as a text accompanying the Seti shadow clock indicates. This situation can be schematized:

| | |
|---|---|
| Hours 1–2: | Re remains outside Nut; the shadow clock has "hand" (*ḏrt*) and lip (*spt*) for these hours. |
| Hour 3: | Re enters Nut; the shadow clock refers to the "tooth" (*nḥdt*), surely of Nut. |
| Hour 9: | Re begins to withdraw from the body of Nut. |
| Hour 10: | Re leaves Nut; the shadow clock text refers to the "vagina" (*k3t*) of Nut. |

100 Hornung, *Nacht und Finsternis im Weltbild der Alten Ägypter* (Tübingen, Ph. D. Dissertation, 1956).

101 *Egyptian Astronomical Texts I*, p. 35.

102 *Grundriss des Laufes der Sterne*, pp. 218–22.

103 *Egyptian Astronomical Texts I*, p. 82.

104 Ibid., pp. 82–83 and 116–18.

Hours 11–12: Re must be outside Nut but not risen; see his location in the shadow clock text at the "thigh" for hour twelve (*mnt*).

It would burden this discussion to delineate each new step in ancient Egyptian timekeeping, but a convenient study is already in press. Hence, the following schematic chart can serve as a useful overview. Here, I have combined the evidence of the shadow clocks of the New Kingdom in conjunction with the later clepsydra. The following quote summarizes the present data:[105]

> The shadow clock recorded in Seti I's cenotaph presents only *four* marks for daylight and not five, operated on the same basis. Of avoiding morning and evening twilight. From the evidence of the schematized or "working" account, there were eight hours in the day that could be roughly measured. In addition, four hours were needed, two of which covered the time before the sun could cast a shadow and two hours after the shadow grew too large to be reckoned. The last line of the text further notes that these remaining two hours of the day immediately precede the night hours, thereby implying that the shadow clock linked up with the night. I agree with Neugebauer and Parker that it is highly plausible that sunrise and sunset formed the dividing line of the first two and the last two. Therefore, the second hour of the day could be easily determined when it began (sunrise) and the twelfth hour's beginning would be likewise dependent upon a simply observed event, sunset. The commencement rather than the end of each day hour was the point of recognition, unlike the system for night reckoning discussed in this paper.

Evidence from a second shadow clock, dated to the reign of Thutmose III can be brought into discussion:[106]

> Because there were ten hour intervals (five times two) measured by five horizontal marks on this apparatus, the divisions that were represented on the Seti I example had now been superseded. The increase in the number

105 Spalinger, "Nut and the Egyptologists."

106 Neugebauer and Parker, *Egyptian Astronomical Texts I*, pp. 117–18. In a private communication, Quack feels that "you would need an enormously long instrument to measure the shadows near sunrise and sunset." Neugebauer and Parker, in fact, argue the same with respect to the Thutmose III example. The shadow clock described in the Seti cenotaph, however, uses four marks and that of Thutmose III had five marks, "thus assuming 6+6=12 hours of daylight." And thus the "division could not be effected by stars but only by a water clock" (*Egyptian Astronomical Texts I*, pp. 118–19).

> of marks on the object — from four to five — implies that this shadow clock measured off ten hours for the day. Therefore, it could only have operated from sunrise to the end of each hour through the eleventh. At the conclusion of the twelfth hour it would have been sunset. To put it another way: the ten castings of a shadow would have covered the ten hours of daytime and the remaining two hours would have covered the extra time following sunrise and preceding sunset. One crucial result of this analysis can now be mentioned. Thutmose's shadow clock implies that the night hours would also have sunrise and sunset as their boundary points. This means that a change had come to pass with regard to a regularized method of time keeping: the night hours no longer were based on the decanal system.

The development that covered the difficulty in determining the beginning of the day is easily set up:

| | DATE | NIGHT HOURS | DAY HOURS |
|---|---|---|---|
| (1) | decanal system | complete darkness | no evidence |
| (2) | Seti I shadow clock | complete darkness assumed | twilight must be included |
| (3) | Amenemhet water clock/[107] Amunhotep III water clock[108] | complete darkness | twilight must be included? [109] |
| (4) | Ramesside star clocks | complete darkness | twilight must be included |
| (5) | Thutmose III shadow clock | measured civil night | civil day |

Here, it can be seen that the development of non-visual parameters, "inventions" or machines if you will, had their effect upon the Egyptian system of timing. As early as the Twelfth Dynasty, the Egyptians had altered the method of sighting the decans.

107 "The Movement of Time. News from the 'Clockmaker' Amenemhet," in Renata Landgráfová and Jana Mynářová (eds.), *Rich and Great. Studies in Honor of Anthony J. Spalinger on the Occasion of his 70th Feast of Thoth* (Prague, 2016), pp. 187–206.

108 The Amunhotep III water clock found at Karnak presents "a situation between 1640 and 1520 BC": Neugebauer and Parker, *Egyptian Astronomical Texts I*, p. 119. Hence, the object can be dated around the time of the reign of Amunhotep I. Mengoli dated the time reckoning on this votive object to "1600 e il 1550 [BC]" in *Astronomia Egizia*, p. 365.

109 Neugebauer and Parker, *Egyptian Astronomical Texts I*, p. 119.

The new system was based upon the transiting of the meridian, a more effective way of placing the location of the star groups. The eastern horizon can be a misleading position due to phenomena that can cause obscurity, such as dust, clouds, and height of the horizon, not to mention the challenge of the first sighting of any star with precision. By moving to the meridian, some of these problems were alleviated. Hence, in ancient Egypt all star clocks would be based upon the celestial meridian and the transit of particular stars.

The decanal star clocks always included the periods of morning and evening twilight. For the day, a separate method of timekeeping was necessary. While data prior to the New Kingdom are rare, certain inscriptions and texts reveal than an irregular system of daytime reckoning was already in use. The previous chart outlines the new parameters introduced by means of the shadow and water clock. The real problem was to mark specific intervals within daytime.

Evidence from these two new artificial means of reckoning daylight time lets us reconstruct the aspects of diurnal reckoning. At first, it appears that shadow clocks told eight hours of the day; morning and evening twilight belonged to the night. Later, in the Eighteenth Dynasty, the water clock was introduced. Scholars assume that shadow clocks were used for the daytime and water clocks for the night, although the latter could easily have functioned in daylight, especially indoors. (As, for example in the speeches of Roman senators.) Both primitive machines witness a change in ancient Egyptian calendrical reckoning. For the first time, artificial means were used to determined the hours. We finally find mention of the hours of the day, perhaps the most important example occurring in the reign of Thutmose III when the pharaoh was fighting against the Asiatic coalition at Megiddo.[110] The older method of the Egyptian daytime revolving around meals, especially those of morning and evening, was slowly disappearing, even if traces of this method can be found throughout native Egyptian history.

Contemporary with the use of shadow clocks for ascertaining the day hours was a further development in telling time during the night. While the so-called Ramesside star clocks are associated with a later period (Dynasties XIX–XX), internal evidence dates them approximately to the reign of Hatshepsut.[111] It may be the case that this newer method — also meridian-based, it can be noted — actually preceded the use of the water clocks for the hours of the night. At any rate, this time the operation was different. Two men sat upon a level surface, undoubtedly the roof of a temple or

110 Parker, "Some Reflections on the Lunar Dates of Thutmose III and Ramesses II," in William K. Simpson and Whitney M. Davis, eds., *Studies in Ancient Egypt, the Aegean, and the Sudan: Essays in Honor of Dows Dunham on the Occasion of his 90th Birthday, June 1, 1980* (Boston, 1981), pp. 146–48.

111 Neugebauer and Parker, *Egyptian Astronomical Texts II: The Ramesside Star Clocks* (Providence and London, 1964).

another, non-religious building. They faced each other. One man noted the position of certain stars on the head, shoulders, and so forth of the second man. Where the specific star groups were located determined the hours of the night. The system remained identical to that of the decans insofar as it was highly inaccurate and would have to have been revised after a certain number of years.

The water clocks, on the other hand, although simplistic in outlook, nevertheless provided greater accuracy. At least in their case one did not have to worry about clouds in the way or the inherent fallibility of the human eye. From the extant material we can observe the relative efficiency which the water clocks brought to Egyptian time keeping. Granted that the civil year was truncated to 360 days, and the flow of water not very scientifically engineered, at least the clocks were portable and so would work anywhere and at any time. And with the switch to clepsydra for the night and shadow clocks for the day, the twenty-four hour diurnal cycle was established. Originally the day comprised only eight hours; but soon, parallel to the system for night time, twelve were ordained for the day.

The institution of sequential hours in a day was a result of the need to ascertain the specific times when religious activities should be performed. Also, secular matters may very well have influenced this development; see the need for nightly watches, for example. One often forgets that most of our written data from ancient Egypt is mortuary in context and religious in outlook. That shadow clocks were important for secular activities is self-evident, and the best example is their use during one king's campaigns. It is nevertheless frustrating that the exact impetus for such spurts in intellectual thought remain unknown to modern researchers. Even the introduction of the first water clock, often claimed to be an Egyptian invention, can be seriously challenged.[112] But was it not the moon that was always connected to the month, whereas the sun was linked to the year?: "I give to you in joy the years of the sun and the months of the moon."[113]

---

112 See the sources listed in note 23, above. To that material we can add J. K. Fotheringham, "The Probable Error of a Water Clock," *The Classical Review* 29 (1915): 236–38. He concluded that the water clocks would be of little use in the early morning. In addition, "we must regard an error of twenty-four minutes an hour as something not uncommon." Fotheringham also proved that these instruments were poor substitutes for sundials, to which I add shadow clocks. Hence, I feel that Parker's contention that the latter were brought along with armies ("Some Reflections on the Lunar Dates of Thutmose III and Ramesses II," pp. 146–47) is given further support.

113 Fritz Hintze et al., *Musawwarat es Sufra: Der Löwentempel. Textband* (Berlin, 1993), p. 245.

## E. PROGNOSTICATIONS ON EXACTITUDE

Important work by the physicist Bradley Schaefer has resulted in a modern and exacting analysis of ancient calculations.[114] With regard to the heliacal rising of Sothis, he has shown that even the Mesopotamian system,[115] arithmetically far in advance of the Egyptian equivalent, was still not reliable. "In all," to quote him, "the observational task of recording the date of the heliacal rising of Sirius has one sigma scatter of a few days."[116] That is to say, naked eye sighting is less reliable than modern scholars believe. The astronomical difficulties with respect to Egyptian lunar dates are also unappreciated. With a very large database not at hand, it was easy for Schaefer to conclude that the crescent visibility predictions are uncertain about twenty percent of the time. Schaefer with LeRoy Doggett concluded that "lunar months based on a few positive sightings from a large number of observers will invariably and mistakenly start early."[117] Therefore, it is unwise to assume that even such fulsome predicative models as P. Carlsberg IX will fulfill our expectations. This difficult situation, however, is better deferred to a later discussion, even though Lord Brougham's famous cookery phrase often applies:

> You have only to take in what you please and leave out what you please; to select your own conditions of time and place; to multiply and divide at

114 Bradley Schaefer, "Astronomy in Historical Studies," *Archaeoastronomy* 14 (1999): 89–108 and "New Methods and Techniques for Historical Astronomy and Archaeoastronomy," *Archaeastronomy* 15 (2000): 121–36 are extremely pertinent to the entire discussion of this study. The difficulties in ascertaining lunar crescent visibility were analyzed by him again in this article on pages 126–29. Needless to say, I have not seen his modern approach reflected in the Egyptological literature. His earlier studies on this subject are "Visibility of the First Lunar Crescent," *Quarterly Journal of the Royal Astronomical Society* 29 (1988): 511–23, and "Lunar Crescent Visibility," *Quarterly Journal of the Royal Astronomical Society* 37 (1996): 759–68.

115 Note in particular Abraham Sachs, "Sirius Dates in Babylonian Astronomical Texts of the Seleucid Period," *JCS* 6 (1952): 105–14.

116 See "The Heliacal Rise of Sirius and Ancient Egyptian Chronology," *Journal for the History of Astronomy* 31 (2000): 149–55, and especially page 151 for the quote. In private correspondence Schaefer maintains that the "first visibility of Sirius on its heliacal rise dates is definitely before sunrise by perhaps half an hour." He adds that this may not imply that the beginning of the Egyptian day is twilight rather than sunrise. "The observation starting a time period," he concludes, "need not be identical with the actual start time of that period."

117 LeRoy E. Doggett and Bradley Schaefer, "Lunar Crescent Visibility," *Icarus* 107 (1994): 402.

> discretion; and you can pay the National Debt in half an hour. Calculation is nothing but cookery.[118]

Despite Greek domination over Egypt, the native civil year still continued, not only in the temples but also on accounts, letters, and the like, in Demotic as well as in Greek. (Whether or not they based their lunar day at dawn and associated it with the disappearance of the lunar crescent is another question.) The real problem with the Egyptian calendrical system is that it did not really approximate the solar year with precision. Hence, the Romans moved over to the Alexandrian year because it had a sixth epagomenal day every four years. But it was effortless for astronomers and students of the ancient world to continue to employ the simple Egyptian Civil Calendar. Ultimately, that is the reason: because it was not mathematically sophisticated but reasonable and easy to work with, this arrangement persisted in Egypt and elsewhere for so long.

118 Quoted by Wells, review of Lynn Rose (see note 7, above), *JNES* 61 (2002): 311.

## APPENDIX: COMMENTS ON THE USE OF LUNAR MONTH LENGTHS IN ABSOLUTE DATING
by Dr. Ronald Wells

In 2002, I published a paper on the use of lunar month lengths in the absolute dating of ancient Egyptian events.[119] This article was based on a presentation that I delivered at the British Museum "Under One Sky" conference held at the end of June 2000. That paper relied heavily on the work of Bradley E. Schaefer[120] and others, but did not include the research of B. D. Yallop, which has since been cited in some Egyptological publications as the best criterion for determining the visibility of the new crescent moon.[121] Yallop's technical note was published in 1997, revised in 1998, while I wrote my "Under One Sky" paper at the end of 1999. Consequently, I did not see his study until much later.

I have been asked to review the recent results on dating the El Lahun series of lunar dates published by Rita Gautschy[122] and on her websites[123] in light of my "Under One Sky" paper and Yallop's visibility algorithm.

First, let me point out that both my study and that of Yallop utilized the extensive research of Schaefer. (See the ten references to Schaefer in Yallop.)[124] Moreover, Yallop was concerned with the determination of first or new crescent visibility because Muslim and other countries base the start of their lunar calendars and certain religious feasts on the timing of this first sighting. His derived algorithm says nothing

119 Ronald Wells, "The Role of Astronomical Techniques in Ancient Egyptian Chronology: The Use of Lunar Month Lengths in Absolute Dating," in Steele and Imhausen, eds., *Under One Sky: Astronomy and Mathematics in the Ancient Near East*, pp. 459–72.

120 Bradley Schaefer (Dept. of Astronomy and Astrophysics of Louisiana State University) is a team member of the Supernova Cosmology Project which discovered "Dark Energy" and the acceleration of the expansion of the universe. The team leader, Saul Perlmutter, was awarded the 2011 Nobel Prize in physics for this group research.

121 B. D. Yallop, *A Method for Predicting the First Sighting of the New Crescent Moon*. (HM Nautical Almanac Office NAO Technical Note 69) (Chilton, June 1997 and updated April 1998).

122 Gautschy, "Lunar and Sothic Data from the Archive of el-Lahun Revisited: Chronology of the Middle Kingdom," in Maarten Horn et al., eds., *Current Research in Egyptology 2010: Proceedings of the Eleventh Annual Symposium* (Oxford and Oakville, 2011), pp. 53–61, and "Monddaten aus dem Archiv von Illahun: Chronologie des Mittleren Reiches," *ZÄS* 138 (2011): 1–19.

123 a. http://www.gautschy.ch/~rita/archast/mond/mondeng.html;
b. http://www.gautschy.ch/~rita/archast/sirius/siriuseng.htm;
c. http://www.gautschy.ch/~rita/archast/ephemeriden.html

124 "A Method for Predicting the First Sighting of the New Crescent Moon."

about the detection of the last crescent moon, which is pertinent to Egypt, because the ancient Egyptians "...began their lunar month on the morning of the day when the old crescent of the moon was no longer visible in the eastern sky before sunrise"[125] — that is, on the day following the sighting of the last crescent, which the ancient Egyptians called *psḏntjw*.

As I pointed out, detection of the last crescent visibility is far more difficult "... than first crescent visibility because the former involves detection of a rising object while for the latter, the Moon is generally 10° or more above the horizon when the sky becomes dark enough after sunset for the crescent to be apparent. Because of the large extinction factor in the Nile Valley on the order of 14 magnitudes at the horizon, it is easier to miss the thin rising crescent which will soon disappear because of impending sunrise."[126] Even for an altitude of 1° above the horizon, the extinction only drops to 7 magnitudes.[127]

In writing about Schaefer's work, Gautschy says: "...he tried to incorporate many parameters such as the transparency of the atmosphere, the site's altitude, the geographical latitude, the temperature, the relative humidity, the aerosol content and the time of the year. He defined a quantity R, which constitutes a logarithmic measure of the Moon's visibility. Schaefer's criterion did not establish itself and today the calculations of the quantity R are no longer available."[128]

She does not explain why it did not "establish" itself, nor why the R-data are no longer available. One reason that Schaefer's work is ignored by Egyptologists is because it invalidates the attempts to determine absolute chronology in ancient Egypt using so-called lunar dates. These researchers fail to consider the observational effects that went into determining R. Failure to "rediscover" or "re-determine" R is no excuse for ignoring the very real quantities that Schaefer built into the parameter.

Yallop, on the other hand, in fact did make such a partial consideration by "...using the modern approach of Schaefer." He stated: "As Schaefer points out, using different extinction coefficients I find an enormous difference between a site at high altitude with a clear dry atmosphere and one at sea level with a humid or dusty atmosphere, so my empirical approach must be confined to a specific type of site *where altitudes above sea level and extinction factors are confined to within narrow limits.*[129] (Italics are mine for emphasis.)

125 Parker, *Calendars*, p. 9.

126 "The Role of Astronomical Techniques in Ancient Egyptian Chronology," p. 463.

127 Ibid., p. 462, for a brief discussion of the large near horizon extinction factor and its computation for ancient Egypt.

128 http://www.gautschy.ch/~rita/archast/mond/mondeng.html.

129 *A Method for Predicting the First Sighting of the New Crescent Moon*, p. 5.

Gautschy certainly recognized that ancient Egyptian observation sites never had "high altitudes," and the atmosphere was usually "humid" and "dusty," probably more so in the past than at present.[130]

Yallop divided his empirical data (his Table 4), taken from Schaefer, into 6 groups or zones defined by a range of values for a visibility factor called *q*, a function of the *arcus visionis* and the topocentric width of the thin lunar crescent. (Note that Yallop made corrections in Schaefer's tables to a number of Julian Date determinations of conjunction, which appear to have been transcriptional errors). The first two *q*-zones, A and B, had values in which the crescent was easily visible (A), or visible under perfect conditions (B). The other 4 groupings of *q* related to: "may need optical aid to detect the crescent," "will need optical aid to determine the crescent," "not visible with a telescope," and "not visible," respectively.

Gautschy chose to use only the first two criteria in her ephemeris computations.[131] She did not distinguish between last and first crescent visibility criteria, also applying the same A–B to the last crescent determinations from the ephemerides. But Yallop's quote above specifically stated that his empirical approach was confined to sites where altitudes above sea level and extinction factors were confined within narrow limits. As noted previously, last crescents are morning events rising from the horizon where the extinction factor is on the order of 7–14 magnitudes, making it difficult to see the thin crescent even for cloudless skies. This amount of extinction was not considered by Yallop, since he was primarily interested in first crescent visibilities, which always occur at elevations where the extinction factor is much smaller.

In attempting to determine an absolute chronology for the El Lahun papyri, Gautschy considered the effects of whether the ancient Egyptians defined the beginning of the day to be at dawn or at sunrise.[132] If the latter, then the previous day would be when the last crescent of the Moon would be visible, since last crescents are always eastern horizon events occurring before sunrise. If the former, then the last crescent would be dated on the morning of the observation. Parker considered it conclusive that the Egyptian day began at dawn.[133] I also gave arguments based on the human circadian rhythm that the day started at dawn.[134] But since the beginning may still be arguable, Gautschy carried through her calculations for both.[135]

---

130 http://www.gautschy.ch/~rita/archast/sirius/siriuseng.htm.

131 http://www.gautschy.ch/~rita/archast/mond/mondeng.html.

132 "Lunar and Sothic Data from the Archive of El Lahun Revisited: Chronology of the Middle Kingdom." See the basic studies concerned with this matter in note 10, above.

133 *Calendars*, p. 10.

134 In my review of Christian Leitz, *Studien zur Ägyptischen Astronomie* (Wiesbaden: Harrassowitz; 1989): *BiOr* 49 (1992): 723–28.

135 "Lunar and Sothic Data from the Archive of El Lahun Revisited."

Gautschy states that "A total of 40 lunar dates can be pinpointed to the first lunar day."[136] Using Yallop's new crescent visibility criteria and several reputable ephemerides programs,[137] she concluded that these 40 dates refer to 1909 BC if sunrise is the definition of the beginning of the day; or 1848 BC if dawn is the definition, a difference of 61 years. She did not pursue the possibility of error limits on the month length sequences from these 40 dates.

However, I would posit the question: Did the ancient Egyptians define "sunrise" as the moment of first gleam, the first sunray that appears on the horizon, or when the lower disk of the sun was tangent to the horizon? There is a difference of one day between the two, disk tangency occurring a day earlier than first gleam at the same point on the horizon in the latitudes of Egypt. I raise the issue of sunrise definition here merely for rigor, to point out that there is an extra day to consider in the determination of where the lunar dates fit. She did not consider this third possibility.

Gautschy then proceeded to use the Sothic rising mentioned in a document dated to the seventh regnal year of an unnamed king, usually considered to be Sesostris III, to attempt to further refine the above dates.[138] She computed that the heliacal rising of Sirius might have occurred between 1882 and 1865 BC.[139] Within that range, she found two solutions with the lunar dates: for dawn, 1872 BC; or for sunrise, 1883 BC, a difference of only 11 years. But the interesting fact here is that she stated that the documents had only 82.5% of their lunar dates fit the first year, while the sunrise definition yielded only a 72.5% fit.[140] Presumably, fits were much poorer for the other years in that range. But the implication is that in the series of 40 El Lahun dates, 7 of them were incorrect on the dawn hypothesis, or 11 were in error in the sunrise model.

This is a very poor determination for an "absolute" astronomical fit. Yet she nevertheless maintains that "...the best fit of lunar data is not necessarily the correct one; this insight has huge implications for the use of such data for chronological purposes."[141] This deduction represents a novel aspect in scientific methodology that I have never come across before: establishing a theory by accepting a predictor that generates such poor fits.

Now let us consider the ability of the *Jmy-r wnwt* priests to make observations of the last crescent accurately. As I noted in my "Under One Sky" paper, Dogget and Schaefer provided capabilities of more than 2500 modern observers in their five

136 Ibid., pp. 54–55.

137 http://www.gautschy.ch/~rita/archast/ephemeriden.html.

138 "Lunar and Sothic Data from the Archive of El Lahun Revisited," "Monddaten aus dem Archiv von Illahun: Chronologie des Mittleren Reiches," and http://www.gautschy.ch/~rita/archast/sirius/siriuseng.htm

139 http://www.gautschy.ch/~rita/archast/sirius/siriuseng.htm.

140 "Lunar and Sothic Data from the Archive of El Lahun Revisited," p. 59.

141 Ibid.

Moonwatch Programs to detect first crescent visibilities.[142] Nearly 800 of these determinations were sorted by the visual acuity, age, and experience of the viewers. In my use of the Doggett and Schaefer tables of their Moonwatch 2 program, I assumed that the two age groups, 30–39 and 40–49, would be representative of a typical *Jmy-r wnwt*. These groups would have made errors of first sightings between 18% and 27% of the time.[143] This capability implies that in a sequence of 12 lunar months of a given year, 2 or 3 of them would have had incorrect starting dates. Schaefer himself used an average figure of 20%.[144]

How does a 27% error translate into the accuracy of the El Lahun documents? It scrambles the sequence of 29–30–30–29 etc. month lengths of the 40 dates in the sequence, since up to 11 of them would have incorrect starts. Therefore, the only way to provide a statistical error limit on the determination of an absolute date fit to these documents would be to start with the month length sequence from them as is, then change one month length according to the alteration of the first month by 1 day and compute the new sequence over again. Then return that month to the original date that the record indicates and alter the next date in the record, etc. After doing this for a 1-day difference, then the whole cycle should be re-computed for a 2-day error, then a 3-day error. One can see that the final determination will have a very wide error range to the extent of invalidating the attempt of an absolute fit completely.

Gautschy did not compute the effects of these priestly errors for either the straight attempts to fit the month lengths to a given year or for those moderated by the Sothic rising. Thus, no real conclusions can be made as to fixing the reigns of the kings involved. Nevertheless, one must credit her with the remaining detail that she put into her analysis, in particular the sine wave charts[145] showing the various periodicities of the Moon, the discussions of the Sothic cycles,[146] and the description of the ephemerides.[147]

There is a strict methodology to being a scientist, aimed at uncovering the truth. Nothing should obscure the pursuit of that truth. The scientific rationale behind a number of modern Egyptologists' persistent attempts to calculate lunar month length sequences as a means of establishing an absolute chronology requires more than a "consensus" of like-minded researchers about the use of astronomical theories. It requires recognition that true scientific methodology means that mathematical

---

142 "The Role of Astronomical Techniques in Ancient Egyptian Chronology," pp. 462–63; and L. E. Doggett and B. E. Schaefer, "Lunar Crescent Visibility."

143 Wells, "The Role of Astronomical Techniques in Ancient Egyptian Chronology," p. 462.

144 "The Heliacal Rise of Sirius and Ancient Egyptian Chronology," pp. 149–55.

145 http://www.gautschy.ch/~rita/archast/mond/mondeng.html.

146 http://www.gautschy.ch/~rita/archast/sirius/siriuseng.htm.

147 http://www.gautschy.ch/~rita/archast/ephemeriden.html.

calculations must include quantifiable error limits based on observational realities, or else incorrect solutions may be the result when models are allowed to exceed those constraints.

In my "Under One Sky" paper I quoted Schaefer's conclusion that "...the current large uncertainties in predicting lunar visibility and in ancient Egyptian procedures do not allow for any possible astronomical solution of Egyptian absolute chronology with lunar dates.[148] A similar conclusion was reached by Peter Huber whom I also cited."[149] He noted in particular that 250 to 300 precisely spaced month lengths would be required to give an accurate determination of an absolute fit. In the 11 years since 2000, little has changed in the use of lunar month length sequences for the determination of an absolute chronology. Certainly, the prediction of the old morning crescent is much more prone to observational errors than that for the new crescent. But also the very likely errors introduced by the *Jmy-r wnwt* priests as observers, which yield incorrect starts of 2–3 lunar months per year, place such wide limits on any given sequence of month lengths that it is exceptionally difficult, if not impossible, to anchor it in any specific year.

148 "The Role of Astronomical Techniques in Ancient Egyptian Chronology," p. 469, referring to Schaefer, "The Heliacal Rise of Sirius and Ancient Egyptian Chronology," p. 154.

149 Ibid., p. 461, note 14.

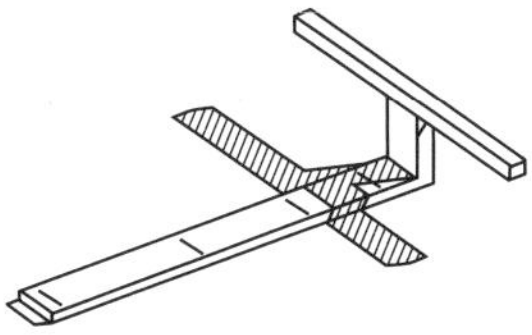

# 2

# Richard Parker and Calendars

The eighty three-page work of Richard Parker, *The Calendars of Ancient Egypt*, is a very peculiar study indeed.[1] Its theme at first appears to be the mechanics of pharaonic and post-pharaonic astronomy and the simpler aspects of this ancient celestial science. Yet it proffers the reader a series of seemingly insurmountable difficulties, not the least of which is the author's sharp, short, and apparently — but not really — concise method of writing. Nothing is left to the imagination, and the dense material can be off-putting. A good reflection of Parker's style is the incessant advance of his ideas without detailed philological explanation. The student of Egyptian language who has not become acquainted with Ptolemaic monumental hieroglyphic writing is immediately perturbed owing to the lack of supportive evidence. With few exceptions, commentary regarding philology or lexicology is brief, even in the footnotes. Rather, Parker expected the reader to master the introduction, wherein he outlined the necessary solar and lunar parameters for dating, calendrics, and astronomical determination. Even there, however, the presentation tended to be lapidary — *knapp*, as the Germans would say. One thus has the feeling, and I am not alone in reflecting upon this aspect of the work, that *The Calendars of Ancient Egypt* resembles, to no small degree, the logical approach used in Bertrand Russell and Albert North Whitehead's *Principia Mathematica*,[2] Ludwig Wittgenstein's *Tractatus Logico-Philosophicus*,[3] or Paul John Frandsen's *Outline of the Late Egyptian Verbal System*.[4]

1 Richard A. Parker, *The Calendars of Ancient Egypt* (Chicago, 1950).

2 Alfred North Whitehead and Bertrand Russell, *Principia Mathematica* (Cambridge, 1910–13).

3 Ludwig Wittgenstein, *Tractatus Logico-Philosophicus*, with an introduction by Bertrand Russell (London, 1922).

4 Paul John Frandsen, *An Outline of the Late Egyptian Verbal System* (Copenhagen, 1974). This work is also included owing to its non-intuitive approach, a methodology in which the preceding two studies excel.

To many, this work represents an insurmountable obstacle to any modern, indeed revisionist, approach to the subject. Therefore, it will be necessary to review Parker's study with an appreciation of the recent scholarly analyses of the same material and thereby to provide a straightforward investigation into ancient Egyptian calendrics. The following detailed remarks are intended to guide the reader beyond Parker but with an appreciation of the seminal nature of his small volume.

In the preface to *The Calendars of Ancient Egypt*, Parker mentioned that he and Harold H. Nelson had undertaken a text volume to supplement the Chicago work, *Medinet Habu III: The Calendar*.[5] This abortive study necessitated a deeper analysis than a mere philological commentary. It is evident from the intense research that went into *Calendars* that Parker soon realized the imponderabilities of the subject. It is clear that Parker was required to investigate calendrics in an effort to perceive the reason why certain festivals occurred on certain days; see his discussion of the lunar dated events of *wꜣgj* or Choiak (*kꜣ-ḥr-kꜣ*), Epiphi, the Feast of the Valley, and so forth.[6] One had to know why these religious celebrations took place. However, Parker was interested in the temporal aspect of these repetitive occurrences and not in the actual unfolding of their religious dramas.[7]

For these reasons, and none other, it was of primary importance to understand the feasts that were not set permanently upon a civil day. A simple contrast could

5 Parker, *Calendars*, Preface. This section indicates that the work was completed at Brown University, Providence, Rhode Island, USA in May 1950. Parker became the first Charles Edwin Wilbour Professor of Egyptology at Brown on 1 July 1948, albeit originally on a part-time basis. In the following year he became a full-time academic there. Cf. I. E. S. Edwards, *From the Pyramids to Tutankhamun: Memoirs of an Egyptologist* (Oxford, 2000), p. 199. Morris L. Bierbrier, *Who Was Who in Egyptology*[4] (London, 2012), pp. 416–77, presents a rather thin bibliography. In the *New York Times* obituary of 5 June 1993, page 28, it is interesting to read that Parker's two important publications, at least as the newspaper reported, were *Calendars* and, with Waldo H. Dubberstein, *Babylonian Chronology 626 BC–AD 45* (Chicago, 1942). When I was studying demotic at Brown University, in answer to one of my queries concerning Waldo Dubberstein, Parker stated that his associate had not been able to find a job in Assyriology and thus left the field. Only later did I find out that Dubberstein's post-1942 career was not as innocuous as what I had assumed. See the article "Indicated Expert on Mideast is Found Dead in Virginia," *The New York Times*, April 30, 1983, page 6.

6 This is important: in order to avoid confusion, I will capitalize the first letter of a month name and not of a feast.

7 The only follow-up studies of Parker of which I am aware that deal exclusively with the matters presented in *Calendars* are: "The Year in Ancient Egypt," *Journal of Calendar Reform* 20 (1955): 81–83; "The Calendar: Past, Present, and Future," *Round Table* 873 (University of Chicago: An NBC Radio Discussion, January 1955): 1–12; and "The Calendars and Chronology," in J. R. Harris, ed., *The Legacy of Egypt*[2] (Oxford: Clarendon; 1971), pp. 13–26.

always be made between the civil New Year's Day, 1 ꜣḫt 1 (I 1) or the fixed Thoth celebration on day nineteen of the first civil month—historically called *Tḫj*[8]—and those others, for example, Renutet or Epiphi. Parker's deep enquiry forced him to retrace the scholarship of Ludwig Borchardt and Heinrich Brugsch, dated and peculiar though their work might seem at first to moderns.[9] This he explicated in a series of intricately woven and tightly interlaced paragraphs, numbered by the quaint symbol §, moving relentlessly forward, step by step, until the naïve reader is lost in the morass of empirical astronomical and calendrical data culled from millennia of native Egyptian history. Parker's Introduction, also short and schematically drawn, is not easy to read, due in part to the figures that accompany the text. They would have been better, I feel, if they could have been three-dimensional, as may be seen in the newer works of Elias J. Bickerman or Robert Hannah.[10] Figures 7 and 8 are not readily accessible to the novice interested in festival dating.

Up to now my reader may have the presentiment that I am complaining too severely. True, for an astronomer such as Ronald Wells or Bradley Schaefer, Parker's overview is simplistic.[11] Yet to an Egyptologist, whether a new or a more mature

---

8 Parker, *Calendars*, p. 75, note 49; Spalinger, "Thoth and the Calendars," in Spalinger, ed., *Revolutions in Time: Studies in Ancient Egyptian Calendrics* (San Antonio, 1994), chapter 3; and "The Date of the *Tḫy* Feast," *SAK* 20 (1993): 289–304.

9 Sir Alan Gardiner had severe, though polite, criticisms of Ludwig Borchardt's lunar orientation. See the study, "Regnal Years and Civil Calendar in Pharaonic Egypt," *JEA* 31 (1945): 11–28. There is no need to extend this analysis, as Borchardt's obsession with the moon is self-evident, especially in his last articles. In 1982, Ronald Wells attempted to sift through the Borchardt material in the Swiss Institute in Cairo. According to the astronomer, Borchardt drew no firm conclusions despite the wealth of information on stellar, solar, and lunar phenomena.

10 I am referring to Elias J. Bickerman, *Chronology of the Ancient World*[2] (Ithaca, 1980), and Robert Hannah, *Greek and Roman Calendars: Constructions of Time in the Classical World* (London, 2005).

11 In light of Bradley Schaefer's remarkable studies, see the significant discussion by Ronald Wells, "The Role of Astronomical Techniques in Ancient Egyptian Chronology: The Use of Lunar Month Lengths in Absolute Dating," in J. M. Steele and A. Imhausen, eds., *Under One Sky: Astronomy and Mathematics in the Ancient Near East* (Münster, 2002), pp. 459–71. Wells used Bradley Schaefer's famous "Moonwatch" program. Later argumentation against Wells' criticism of empirical or naked-eye sighting cannot fault the scientific principles enunciated by him. Therefore, I alert the reader to these studies by Schaefer: "Astronomy and the Limits of Vision," *Vistas in Astronomy* 36 (1993): 311–61, "Lunar Crescent Visibility," *Quarterly Journal of the Royal Astronomical Society* 37 (1996): 759–68, "The Heliacal Rise of Sirius and Ancient Egyptian Chronology," *Journal for the History of Astronomy* 31 (2000): 149–55, and with William Hillier, "Refraction Near the Horizon," *Publications of the Astronomical Society of the Pacific* 102 (2000): 796–805.

scholar, such is not the case. Consider, for example, William Murnane's comments on Parker's study or, equally, the useful chapter of Peter der Manuelian on Amunhotep II's so-called "co-regency."[12] Both Egyptologists, whose scholarship no one would fault, came up with incorrect conclusions based upon Parker's 1950 research. The 1985 volume of Rolf Krauss is perhaps the only one-volume compendium dealing with these matters.[13]

Parker, one must realize, never identified his main theme at the start. If he had, then perhaps the novice reader would have seen why he spent a large amount of time hypothesizing three separate lunar calendars. This point, the true substructure and skeleton of the volume, is what concerns us in this presentation. At the beginning, the reader is faced by a demotic manuscript, P. Carlsberg IX, which was first studied in detail by Neugebauer. Parker brilliantly reconstructed it as a unity.[14] Why,

---

12 William Murnane, "Once Again the Dates For Thutmose III and Amenhotep III," *JANES* 3 (1970/71): 1–7; and Peter der Manuelian, *Studies in the Reign of Amenophis II* (Hildesheim, 1987), pp. 1–19. With respect to the "coregency" of Thutmose III and Amunhotep II, der Manuelian followed Parker's chronological set-up. That conclusion, however, was dependent upon a fixed absolute date, one ordered within Parker's study, "The Lunar Dates of Thutmose III and Ramesses II," *JNES* 16 (1957): 39–43, where the accession of Thutmose was placed in 1490 BC. Yet, with Edward Wente, "Thutmose III's Accession and the Beginning of the New Kingdom," *JNES* 34 (1975): 265–72, this was apparently refuted, and the year set lower at 1279 BC. If the latter is followed, then Parker's calculations concerning the "coregency" do not hold. I presume that this is the reason why Erik Hornung, "The New Kingdom," in Erik Hornung, Rolf Krauss, and David Warburton, eds., *Ancient Egyptian Chronology* (Leiden and Boston, 2006), p. 203, saw difficulties in der Manuelian's argumentation. Parker's seminal work on Thutmose III and Amunhotep II is "Once Again the Coregency of Thutmose III and Amenhotep II," in E. B. Hauser, ed., *Studies in Honor of John A. Wilson* (Chicago: 1969), pp. 75–82.

13 Rolf Krauss, *Sothis- und Monddaten: Studien zur astronomischen und technischen Chronologie Altägyptens* (Hildesheim, 1985). I have also consulted his Dissertation, written at the Freie Universität zu Berlin, *Probleme des altägyptischen Kalenders und der Chronologie des Mittleren und Neuen Reiches in Ägypten* (Berlin, Ph.D. Dissertation, 1981).

14 *Calendars*, Chapter 1. See also Wells, "Some Astronomical Reflections on Parker's Contributions to Egyptian Chronology," in Leonard H. Lesko, ed., *Egyptological Studies in Honor of Richard A. Parker* (Hanover and London, 1986), pp. 165–71; Krauss, *Sothis- und Monddaten*, pp. 278–79; Spalinger, "Under the Moon of Earth," in Spalinger, ed., *Revolutions in Time*, pp. 72–78, with "Calendrical Comments," *BiOr* 51 (1994): 5–20; Leo Depuydt, "The Demotic Mathematical Astronomical Papyrus Carlsberg 9 Reinterpreted," in W. Clarysse, A. Schoors, and H. Willems, eds., *Egyptian Religion: The Last Thousand Years. Studies Dedicated to the Memory of Jan Quaegebeur* II (Leuven, 1998), pp. 1277–97; and Chris Bennett, "Egyptian Lunar Dates and Temple Service

however, did he use the cycle of twenty-five years and claim that it is Egyptian in origin when no evidence exists to indicate that the ancient Egyptians ever worked with arithmetical cycles? The Neo-Assyrians and, more successfully, the Babylonians did, but nowhere predating P. Carlsberg IX do we know that the inhabitants of the Nile Valley or the Delta ever constructed a mathematical formula of a cyclical nature.

However, it may be best to sidestep this issue, as it would lead us into the interstices of cultural borrowings. Parker wished to present, in a simple fashion, a twenty-five year template based upon original sightings of the moon. P. Carlsberg IX worked accurately in order to establish an artificial system of months, each having either twenty-nine or thirty days.[15] Parker therefore maintained that this papyrus

Months," *BiOr* 65 (2008): 525–54. In my detailed review of *BiOr* 51 (1994): 5–20, I brought into consideration the little-known 25-year cycle in P. Ryl. Inv. 666: Eric G. Turner and Otto Neugebauer, "Gymnasium Debts and New Moons," *BJRL* 32 (1949): 80–96. In that fragmentary papyrus one can see that the cycle has to be dated to -180 to -179. Hence, it is considerably earlier than the date of P. Carlsberg IX, which is Tiberius 6, Thoth 1. Depuydt's novel analysis is worth considering. However, in light of Bradley Schaefer's twenty or so years of research on the weakness of utilizing lunar dates, his treatment is open to criticism. Sandra L. Lippert, "*Au clair de la lune* - The Organisation of Cultic Service by Moon Calendar in Soknopaiou Nesos," in *Actes du IXe Congrès International des Études Démotiques; Paris, 31 août – 3 septembre 2005* (Cairo, 2009), pp. 183–94, also attempted, at least at the start, to place her given empirical data into the girdle of P. Carlsberg IX, as Depuydt did. Cf. Lippert's comments on page 189: "Parker, and after him, Depuydt, reconstructed from this a whole table, but, as Depuydt points out, his reconstruction of the P. Carlsberg IX table is *not* [my italics; this is the nub of the matter — Depuydt's restoration is one of many] the only possible arrangement of lunar months within the 25-year cycle." We are thus left with wondering what, exactly, was the lunar basis for the Carlsberg papyrus. Was it, in fact, Egyptian or otherwise? Because Lippert's research seems to have disproved the possibility of adapting the paradigm to the cultic service at Soknopaiou Nesos, we are forced back to the (expected) conclusion that if a lunar organization was in operation, it depended upon naked-eye sightings. Add the background data presented by Bennett, *Alexandria and the Moon. An Investigation into the Lunar Macedonian Calendar of Ptolemaic Egypt* (Leuven, 2011).

15 I have discussed the problem of allocating, or assigning, this schematic system to the Egyptians in "Calendrical Comments," *BiOr* 51 (1994): 5–20. I then preferred to see it enter Egypt from the east and in particular during the Persian Empire, but I argued for a Babylonian origin. This I no longer do. The comments of Neugebauer are worthwhile to cite here: "The fundamental idea of *cyclical* intercalation, however, which is so pronounced in the Seleucid system, has no counterpart in the Athenian Procedure" — "The 'Metonic Cycle' in Babylonian Astronomy. Studies in Ancient Astronomy. VI.," in M. F. Ashley-Montagu, ed., *Studies and Essays in the History of Science and Learning Offered in Homage to George Sarton. On the Occasion of his Sixtieth Birthday, 31 August*

proved that the Egyptians produced "the first schematic calendar" in the fourth century BC. He then purposely underlined — this was the era of typewriters, it must be noted — the following passage: "this lunar calendar cannot have existed before the civil calendar was introduced."

I always felt that chapter 2 of *Calendars* was misplaced and, properly speaking, should have been located at the end of the book. Notwithstanding this, the reader will now see why Parker gravitated to P. Carlsberg IX. Based upon it, he set up an artificial lunar based system which was functioning (many centuries) after the invention of the Civil Calendar. Hence, the argument revolved around another scheme for/of time reckoning, too well-known to Egyptologists for anyone in 1950 to be disturbed over it. In the following chapter, therefore, Parker threw his sights back in time to a date preceding the introduction of the Civil Calendar. He surveyed past scholarship, making, for the most part, effective rebuttals against previous material. Then he commenced to reconstruct the original or primitive annual calendar of the Egyptians.[16] Mindful of Neugebauer's unsuccessful attempt at defining the original, Parker set about reconstructing the original Civil Calendar of the Egyptians. He began to emphasize Sothis and link it with the moon.[17]

---

*1944* (New York, 1946), pp. 446–47; the whole chapter runs from pages 435–47. I would like to add my position that cyclical procedures were extremely uncommon, but not impossible, to the Egyptians. See notes 14–15 in Chapter 3.

16 On the cycle of Sothis, the later so-called Phoenix year, see Oleg D. Berlev, "Precisely Two Sothic Periods Between Year 18 of King *Sénw*/Tosorthros and Year 2 of Pharaoh Antoninus Pius," in O. I. Pavlova, ed., *Ancient Egypt: Language, Culture, Consciousness* (Moscow, 1999), pp. 42–62. He turned to Parker's later Sothic analysis, "The Sothic Dating of the Twelfth and Eighteenth Dynasties," in Janet H. Johnson and Edward F. Wente, eds., *Studies in Honor of George R. Hughes: January 12, 1977* (Chicago, 1976), pp. 177–89. The study is also summarized in Berlev's review of Peter der Manuelian ed., *Studies in Honor of William Kelly Simpson* (Boston, 1996), in *BiOr* 56 (1999): 600–601. It cannot be left unsaid that on page 31 (§ 153) of his *Calendars*, Parker referred to the Loango (west coast of Africa), who also used a lunar calendar based on Sothis and just possibly, he felt, may have influenced the Egyptians. This provoked Gardiner, of course ("The Problem of the Month-Names," *RdE* 10 [1955]: 22), but the importance of the Loango was earlier indicated by two books which Parker consulted: Hutton Webster, *Rest Days: A Study in Early Law and Morality* (New York, 1916), especially page 177, note 1; and Martin Nilsson, *Primitive Time Reckoning: A Study in the Origins and First Development of the Art of Counting Time among the Primitive and Early Culture Peoples* (Lund, 1920).

17 Neugebauer had recruited Parker to Brown University as he later recruited Abraham Sachs, previously also at the University of Chicago. He wanted to work with them and together to publish material on ancient natural science. There is no doubt that Parker was "headhunted" in the late 1940s when he was still at Chicago. Indeed, Parker's earlier book with Waldo H. Dubberstein, *Babylonian Chronology 626 BC–AD 45*, as well as

In essence, it was the heliacal rising of Sothis that determined the first day of the lunar-based year.[18] (This was *wp rnpt.*) Parker's figure 14 on page 31 succinctly outlined his position. But he added one important factor not known from any Egyptian text. This was the intercalary month.[19] Parker claimed that this extra month was necessary in the original or first lunar calendar. It was not. But then he downplayed the innovation by immediately proceeding to a discussion of the religious celebrations of bygone ages. He assumed the equivalence between *prt Spdt*— the Egyptian term for the heliacal rising of Sothis — and *wp rnpt* and forged ahead by dealing with other religious celebrations of bygone ages. Suddenly, however, he covered the exacerbating issue of *tpj rnpt.*

Passing over the matter of proofs, Parker concentrated upon these feasts in an effort to (1) determine the nature of the *wꜣgj* feast, number four in the Old Kingdom feast list,[20] and (2) work out chronologically the Old Kingdom feast list in which *tpj rnpt* occupied place three; the all-important Sokar celebration the fifth; the event of "Heat"[21] the seventh; and the Min festival the eighth.

---

his solo work showed excellence in the areas of calendrics, astronomy, Egyptology, and Assyriology. Let me signal out the following three "early" publications of Parker that bear upon his subsequent intense studies of the Egyptian calendar: "Darius and His Egyptian Campaign," *AJSL* 58 (1941): 373–77, "Persian and Egyptian Chronology," *AJSL* 58 (1941): 285–301, and "Ancient Jewish Calendation: A Criticism," *JBL* 63 (1944): 173–76. There is no doubt that this research was partly inspired by the then Chicago scholar George Cameron; cf. the latter's "Darius, Egypt and 'The Lands Beyond the Sea,'" *JNES* 2 (1943): 307–13, and "Darius and Xerxes in Babylonia," *AJSL* 58 (1941): 314–25.

18 Most recently, see Teije de Jong, "The Heliacal Rising of Sothis," in Hornung, Krauss, and Warburton, eds., *Ancient Egyptian Chronology*, pp. 432–38; with Krauss, "Egyptian Sirius/Sothic Dates, and the Question of the Sothis-Based Lunar Calendar," in the same volume, pp. 439–57 and especially 450–52. Schaefer's "The Heliacal Rise of Sirius and Ancient Egyptian Chronology" is not referred to. But we can add Ulrich Luft, "Illahunstudien IV: zur chronologischen Verwertbarkeit des Sothisdatum," *SAK* 16 (1989): 217–33.

19 The existence rather than the mere possibility of an intercalary month is stated in *Calendars*, p. 26, § 126: "Whenever the first day of lunar Thoth [he means lunar *wp rnpt*] would fall before the first day of civil Thoth [originally civil *Ṯḫj*— my comment], the month is intercalary." Parker underlined this passage.

20 See Spalinger, *The Private Feast Lists of Ancient Egypt* (Wiesbaden, 1996). The recent study of Katherine Eaton, "Monthly Lunar Festivals in the Mortuary Realm: Historical Patterns and Symbolic Motifs," *JNES* 70 (2011): 229–45, is connected to the older series of lunar day festivals. The author provides additional data concerning the early lunar festivals in Egypt. On the other hand, no exacting historical calendrical analysis is given.

21 It should be understood that originally there was probably one long "month" or a short "season": see Martin Nilsson, *Primitive Time-Reckoning*, chapter 7; Spalinger, "Notes on Ancient Egyptian Calendars," *Or* 64 (1995): 17–32; Parker, *Calendars*, pp. 46 and 77, note 100.

This is complicated work and demands a careful reading that can be absorbed only very slowly by the outsider. One's attention is liable to flag at this point and others. Parker was the first scholar, following the primitive analyses of Brugsch and the later all-too-complicated ones of Borchardt, to provide a clear-cut organizing principle behind the Old Kingdom feast lists.[22] Yet Parker was caught, as was Borchardt earlier, in the *wꜣgj* problem: two of them could occur within a single civil year. The civilly based one was fixed, as all knew and expected, on day eighteen of civil month one; however, there was a "moveable" *wꜣgj* as well, first described in detail by Borchardt.

Here is the heart of the matter. For the Egyptians, one civil *wꜣgj* was not enough.[23] The second had to be lunar-based. If so, it would have operated according to the earlier or first lunar calendar. Considering the evidence from other cultures, no other type of calendar could have preceded it, and of course that includes the Civil Calendar in which the fixed *wꜣgj* feast was permanently set on I *ꜣḫt* 18. The last was created on a schematic basis after the original lunar calendar, which had not worked successfully as a method of temporal reckoning.

The related question arises: when did the replacement calendar, the civil one, come into being? Additional data culled by Parker from Borchardt et al. indicates that some Egyptian festivals were still regulated by the moon during the heyday of pharaonic Egypt. Up to this point, Parker had avoided the problem of any intercalary month. His Illahun data had not yet been properly published and therefore could not explain his assumption. Likewise, Parker's attempt to enlist disparate source material such as the two-dimensional representations of water clocks revealed nothing extraordinary.[24] Quite to the contrary, the astronomical ceilings of, say, the Ramesseum or Medinet Habu indicate that the Egyptian Civil Calendar ran the time. Nowhere—indeed at no time—was there any evidence that a certain *Ḏḥwtjt*, or Thoth, was an intercalary month. If the original lunar calendar existed, and no one would doubt that assertion, then it apparently lacked an extra month, call it XIII. Here, it should be recognized that we do not know if there were twelve months in a normal ("short") Egyptian lunar year. Parker, in fact, seems to have opted for eleven.

Chapter 3 proceeds in a most remarkable way by providing a wealth of empirical data from P. Ebers with a useful, if imperfect, discussion of water clocks. Additional

---

22 *Calendars*, pp. 34–36.

23 The study of Luft, "The Date of the *Wꜣgy* Feast: Considerations on the Chronology of the Old Kingdom," in Spalinger, ed., *Revolutions in Time*, pp. 39–44, remains standard. Krauss, "Wenn und Aber: Das Wag-Fest und die Chronologie des Alten Reiches," *GM* 162 (1998): 43–63 attempted a refutation of Luft. For additional data see Hana Vymazlová, "Some Remarks on the *wꜣg*-Festival in the Papyrus Archive of Raneferef," in Hana Vymazlová and Miroslav Bárta, eds., *Chronology and Archaeology in Ancient Egypt (The Third Millennium B.C.)* (Prague, 2008), pp. 137–43.

24 Spalinger, Month Representations," *CdE* 70 (1995): 110–22. In this case Parker was inaccurate because the water clocks had to have been set within the civil calendar.

Fig. 2 Parker's Names of the Months (reproduced after Parker, *The Calendars of Ancient Egypt* [Chicago: University of Chicago; 1950], p. 45).

| Lunar Calendar Month Names | | | Civil Calendar Month Names | |
|---|---|---|---|---|
| Intercalary | | *Ḏḥwtyt* | Early | Late |
| I | *ꜣḫt* | *tḫy* | *Ḏḥwty* | Thoth |
| II | " | *mnḫt* | *p-nꜣIpt* | Phaophi |
| III | " | *Ḥt-ḥr* | *Ḥt-ḥr* | Athyr |
| IIII | " | *kꜣ ḥr kꜣ* | *kꜣ ḥr kꜣ* | Choiak |
| I | *prt* | *šf bdt* | *tꜣ ꜥꜣbt* | Tybi |
| | | | *pꜣ ḫnw Mwt** | |
| II | " | *rkḥ wr* | *p-n mḫr* | Mechir |
| III | " | *rkḥ nḏś* | *p-n ꜣImnḥtp* | Phamenoth |
| IIII | " | *Rnwtt* | *p-n Rnwtt* | Pharmuthi |
| I | *šmw* | *Ḫnśw* | *p-n Ḫnsw* | Pachons |
| II | " | *Ḫnt-ḫty* | *p-n ỉnt* | Payni |
| III | " | *ꜣIpt ḥmt* | *ỉpỉp* | Epiphi |
| IIII | " | *wp rnpt* | (*mśwt Rꜥ-ḥr-ꜣḫty*)** | Mesore |
| | | *Rꜥ-ḥr-ꜣḫty* | *wp rnpt* | |
| | | | *pꜣ šmt n Ḥr**** | |

* The more usual early name; cf. Černý, *Ann. Serv.*, XLIII, 175.
** This is not yet attested as a month name in the 20th dynasty. The reference in Černý is to the feast on I *ꜣḫt* 1and not to III *šmw*. See § 236 later.
*** The probable name for III *šmw* in ostr. Brit. Mus. 5639a, 4.

source material comes from the famous Tanis Geographical Papyrus, Ptolematic Period data concerning foundation building, the Senenmut ceiling, and so forth. By this time there is no unity in the discussion. If one is well-versed in the background data, then all is secure and understood. Otherwise, the outsider will have great difficulty in attempting to digest the material.

To be fair, this always must be a hallmark of scientific investigations — namely, the writer's assumptions concerning the necessary intellectual qualifications of the reader. Moon chronology, for this is what Parker aimed at, requires astronomical insight and some technical arithmetical processing ability. (Indeed, as Neugebauer showed, higher-level arithmetical competence could also be necessary, but that was with the Babylonian astronomical data and it is not needed for the considerably simpler systems prevalent in ancient Egypt.) Parker does attempt to buttress his festival dates. But are his efforts recognizable as such to interested and engaged readers? I have always wondered if any of his perceptions were carried over from this detailed and erudite exposition into the minds of his readership who lacked, or presently lack, careful mathematical training.

Take into consideration his analysis of the names of the Egyptian months. At this point he was approaching the crux of the exposition. In Table 7, Parker reconstructed the original lunar names of the Egyptian months and their later civil equivalences. Unfortunately, the detritus of his assumed intercalary month remained. Note, however, that the "rule" X > X-1 applies. By this I mean, and it was Parker who had great perspicacity to recognize the situation, the name of the later civil months could be associated with the names in the first lunar calendar by a simple arithmetical formula. Subtract the order; e.g., V-1 = IV. Here, then, is the second important "discovery" or, if it is preferred, "analysis" of the civil month names. See the list of Parker in figure 2.

Parker observed that the eponymous feasts did not fit the names of the civil months but retained the order of the lunar month names when the newer Civil Calendar came into being, with one caveat, that being the simple rule Y = X-1. (To be scrupulously fair, Parker did not word his conclusions this way.) The three key festivals that concerned him, and which are best to analyze, are as follows:

1) *Wp rnpt* = name of civil month XII, *wp rnpt* = day one of the first civil month,
2) *Kꜣ-ḥr-kꜣ* is the name of civil month IV, but the *kꜣ-ḥr-kꜣ* festival took place in civil month V.[25]
3) The celebration of *rnwtt* occurred in civil month IX, yet it is the name of civil month VIII.

---

25 "Peculiar" writings of *kꜣ-ḥr-kꜣ* may be found in the following cases. Note that I am transcribing only the signs that are definite, but with *kꜣ-ḥr*, is *kꜣ-ḥr<-kꜣ>* to be preferred?
a) O. DEM 1088 recto 6 (seen) — … *kꜣ-ḥb*
b) O. DEM 1245 col. I, 11 — *kꜣ-ḥb-kꜣ* ...
c) O. DEM 1059 recto 7-8 — *kꜣj-hb-kꜣ*
d) Cairo Calendar Verso XIV 4: Černý, "The Origin of the Name of the Month Tybi," *ASAE* 43 (1943): 179; Neugebauer and Parker, *Egyptian Astronomical Texts I. The Early Decans* (London, 1960), pp. 119–20 (the best discussion); Bakir, Abd el-Mohsen, *The Cairo Calendar No. 86637* (Cairo, 1966), Pls. XLIV–V and XLIVA; and Christian Leitz, *Tagewählerei: Das Buch ḥꜣt nḥḥ pḥ.wy ḏt und verwandte Texte.* Tafelband (Wiesbaden, 1994), pl. 44 — *kꜣ-ḥr-[ḥ]b*
e) Berlin P. 12635 recto 8 (seen) — *kꜣ-ḥr*
f) Tomb of Amunhotep: *Urk.* IV 107.7 — *kꜣ-ḥb*
g) P. Brussels E 5043: Louis Speleers, *Le papyrus de Nefer Renpet: un Livre des morts de la XVIIIe Dynastie aux Musées royaux du cinquantenaire à Bruxelles* (Brussels, 1917), p. 37 and pls. XIX and XXVI; with Gardiner, "The Problem of the Month-Names," — *kꜣj-ḥr-kꜣ*
h) Tomb of Paheri: Gardiner, ibid., 15 note 1, reinforced by Černý's notes in his copy of the volume which I have seen; Joseph John Tylor, *Wall Drawings and Monuments of El Kab: The Tomb of Paheri* (London, 1895), pl. 4; *Urk.* IV 109.15 is somewhat incorrect,

It ought to be evident that the ground was not properly cleared, despite Parker's claim.[26] The confusion inherent in his analysis of the data was compounded by his later research, when Parker was under enormous pressure by Sir Alan Gardiner to re-evaluate the material. Therefore, in his 1950 publication Parker addressed Gardiner's futile 1906 discussion of Mesore, the later name for the Egyptian twelfth civil month.[27] The issue is obvious: Coptic calls the last civil month by this name. Surely, Mesore indicates the "birth" of the sun god. The name is essentially equivalent to Ramesside *Mswt Rꜥ*, although we also find the designation of *Pꜣ šm n Ḥr* for the same thirty-day interval.[28] At the crucial point, Parker had recourse to the Tanis Geographical Papyrus, which may reflect an Eighteenth Dynasty original.[29] In the first section that deals with a water clock, the designation of *Wp rnpt* appears for civil month XII. While there is nothing extraordinary so far, the source also contains the language "Re sails north" or "the sailing of Re north." Once more we are witnessing, from different perspectives to be sure, the "rise" of the sun god in connection with the final civil month of the Egyptian year. When Re goes north, he is increasing in power or strength, becoming more mature, and so forth. A new cycle has begun. Re is likewise

as Gardiner saw very early in his career: Nina de Garis Davies and Alan H. Gardiner, *The Tomb of Amenemhet (No. 82)* (London, 1915), p. 97, note 3; and now Spalinger, "The Date of Amunhotep II's First Accession," *SAK* 40 (2011): 387–97 — *kꜣ-<ḥr>-kꜣ*

i) Senenmut ceiling: Parker, *Calendars*, pl. I; and Neugebauer and Parker, *Egyptian Astronomical Texts III: Decans, Planets, Constellations, and Zodiacs*, Plates (London, 1969), pl. I — *kꜣ-kꜣ*

j) P. BM 10,335 verso 2: Warren R. Dawson, "An Oracle papyrus B.M. 10335," *JEA* 11 (1925): pl. XXXVII — *kꜣ-ḥr*

k) O. Berlin 12406 recto 11, verso 1 (seen) — *kꜣ-ḥr*.

26 *Calendars*, p. 46.

27 Gardiner, "Mesore as the First Month of the Egyptian Year," *ZÄS* 43 (1906): 136–44. His position has been taken to a more intricate level by Leo Depuydt, *Civil Calendar and Lunar Calendar in Ancient Egypt* (Leuven, 1997). Gardiner, nonetheless, knew that Mesore *ought* to refer to civil month I and not XII. We now know why he was on the right track. Note the Leitz review of Depuydt's volume in *BiOr* 57 (2000): 75–81.

28 Three New Kingdom (Ramesside) cases of the name *šm n Ḥr* for civil month XII are known to me: O. BM 29560 (formerly 5639a — Parker, *Calendars*, p. 45, last note on bottom), O. Gardiner 141 (seen), and P. DEM 2 (verso) — *KRI* VI 260.8. Cf. R. van Walsem, "Month-Names and Feasts at Deir el-Medîna," in R. J. Demarée and Jac. J. Janssen, eds., *Gleanings from Deir el-Medîna* (Leiden, 1982), pp. 220, no. 23, and 228, nos. 107–8.

29 Francis Ll. Griffith, *Two Hieroglyphic Papyri from Tanis* (London, 1889), pl. IX (Part I, 2) as discussed in Parker, *Calendars*, p. 41. The key passage was misunderstood by Adelheid Schlott-Schwab, *Die Ausmasse Ägyptens nach altägyptischen Texten* (Wiesbaden, 1981), p. 47, note 192.

FIG. 3 Parker's Evidence for the Name of the Last Civil Month (taken from Griffith, *Two Hieroglyphic Papyri from Tanis* [London: Trübner; 1889], pl. IX, top right).

"born"; hence, Mesore. The concepts are identical and, in fact, coincide *in thought* with the older designation for civil month XII, "The year is opened," *Wp rnpt.*[30] One very late case is present in the "Tanis Geographical Papyrus" (fig. 3).

One does not have to make any computational alterations to fit the evidence. Sometime in the New Kingdom — Gardiner and Depuydt would argue for the Nineteenth–Twentieth Dynasty — the Egyptians gave a new name to their twelfth civil month.[31] Nothing more need to be said with regard to either of their unprovable hypotheses. Contortions of chronology can be avoided. My simple answer provides a solution to the data that confounded Parker. Of course, he wanted *Ḏḥwtjt* to be the original name of the hypothesized intercalary month in his "first lunar calendar,"

30 I feel that the title is a quote from a performative act. Most certainly, the verb is not an infinitive. I understand it as a passive *sḏm=f*. We might also read: "That which opens the year."

31 Depuydt, *Civil Calendar and Lunar Calendar in Ancient Egypt*, and Gardiner, "Mesore as the First Month of the Egyptian Year."

which was subsequently transferred over to the name of civil month I. However, in this instance he was confusing the information which he had so carefully presented earlier: namely, lunar month XII was *not Wp rnpt*. How could it be? *Wp rnpt* opens, inaugurates, or commences a cycle, a span of time. Properly speaking, it should refer to the beginning of the year, whatever year that was. Hence, *Wp rnpt* was the original name of lunar month I and was transferred into the subsequent civil calendar as month XII.

According to basic observation, the civil month names make sense. The reason why civil month IV was labeled *Kꜣ-ḥr-kꜣ* was that its lunar predecessor was lunar month V. The same can be said for lunar *Rnwtt*. At first, it was month IX, and in the newly adopted civil calendar, month VIII. For this reason we have no difficulty in understanding the dates of the feasts associated with these month names. Originally:

1) The feast of *wp rnpt* took place in lunar month I and not in lunar month XII. With the transference of month names, *Wp rnpt* was then applied to civil month XII.
2) The feast of *kꜣ-ḥr-kꜣ* took place in lunar month V, but with the transference of month names *kꜣ-ḥr-kꜣ* was then applied to civil month IV.
3) The festival of Ernouthis, or *rnwtt*, still occurred in civil month IX in historical times; but the name of civil month VIII was Pharmuthi, obviously derived from *rnwtt*, the important religious event.
4) The civil month Hathor, III, has a name obviously derived from the eponymous feast *Ḥwt-Ḥr* which took place in the fourth civil month.
5) The feast of *rkḥ wr* is once located on day one of civil month VII, but the name of civil month VI was *rkḥ wr*.[32]

These are five excellent cases from which to draw conclusions diametrically opposed to those of recent calendrically oriented scholars.[33] They do not depend upon any fancy or complicated mathematical adjustments. In particular, note that I have cho-

32 Thomas Garnet Henry James, "The date of the month *rkḥ wr*," *JEA* 41 (1955): 123. He proved his case. I have seen the original in Berlin as well as the Eugène Dévaud copy. The basic study remains, nonetheless, that of Luft, *Die chronologische Fixierung des ägyptischen Mittleren Reiches nach dem Tempelarchiv von Illahun* (Vienna, 1992), pp. 82–83. There are minute differences between Dévaud's transcription and that of Luft. Despite some hesitations on the part of James, it is self-evident that the date of III *prt* 1 refers to the feast of *rkḥ wr*. That is to say, the latter is not a month name. See my detailed comments in Chapter 3, note 48.

33 I am adumbrating my earlier discussion, "The Beginning of the Civil Calendar," in Miroslav Bárta, Filip Coppens, and Jaromír Krejčí, eds., *Abusir and Saqqara in the Year 2010* (Prague, 2012), pp. 723–35. This shall be discussed more in the following chapter.

sen the simplest cases. However, we *may* still regard the original lunar year as having been composed of three seasons that are well-known from the later Civil Calendar: Inundation (*ꜣḫt*), Emergence (*prt*), and Harvest (*šmw*).

Now we can address the Civil Calendar of Parker. So far, we have seen that his assumptions were few though significant, even if to Egyptologists they appeared recondite. First, the Civil Calendar, which he set rather early, was based upon the twelve–thirteen month system of a hypothesized lunar year. Second, the names of the months in the Civil Calendar were derived from the earlier names of the lunar months. Third and last, the Civil Calendar had to be set by one celestial or one earthly phenomenon. It is the third supposition that concerns us at this point. Parker knew that the 1938 hypothesis of Neugebauer concerning the inauguration of the Civil Calendar was not acceptable.[34] The argument of Neugebauer centered upon the establishment of a ca. 365-day internal cycle which would have been worked out based upon the annual flow of the Nile; i.e., inundation. This presupposition was effectively rejected by Robert Walter Sloley.[35] Hence, Parker logically concentrated upon lunar events and concluded that the Civil Calendar was *not* tied to Sothis at its inception but instead to a variable yearly heavenly phenomenon, the moon. His "proposed solution," one that he ignored in his later research, was questioned by Gardiner and seems not to have been followed by most Egyptologists. This was a correct reaction. Yet there is little doubt that Parker's monumental and paradigmatic volume of 1950 pushed lunar sightings to the fore and had the advantage of avoiding the pixilated and "hyper-rational" approach of Borchardt. As a result, Parker is easier to understand than Borchardt.

But Parker's methodology slips here, and it is notable that in chapter 4 of *Calendars* the arithmetical averaging of lunar years meets with the same criticism by Sloley that Neugebauer encountered.[36] Moreover, there is no reason to reject Sothis as the determinate factor in the creation of the first civil day, after explaining the importance of this star for Egyptian civilization, calendrics, and astronomy. I see the difficulties that a lunar or civil year, without any control, would have entailed, owing to the steady shift in the year over a relatively short period of time. Parker did too. That is why the attempted reconciliation of the new civil year with the older lunar year meant so much to him. If one takes a close look at figure 20, the "solution" to the feasts that Parker had already discussed is revealed. This is his procedure:

34 Otto Neugebauer, "Die Bedeutungslosigkeit der 'Sothisperiode' für die älteste ägyptische Chronologie, *Acta Orientalia* 17 (1938): 159–95. See, however, Parker's criticism of this work in *Calendars*, pp. 78–79, notes 2 and 7.

35 This was brilliantly presented by Robert Walter Sloley in "The Origin of the 365-Day Egyptian Calendar," *ASAE* 48 (1948): 261–65. Neugebauer never reacted to the rebuttal. The argumentation of Sloley is clever and sly; it is also correct.

36 *Calendars*, p. 53.

Fig. 4 Parker's Origins of the Hypothetical "Second Lunar Calendar" (reproduced after Parker, *The Calendars of Ancient Egypt*, p. 54).

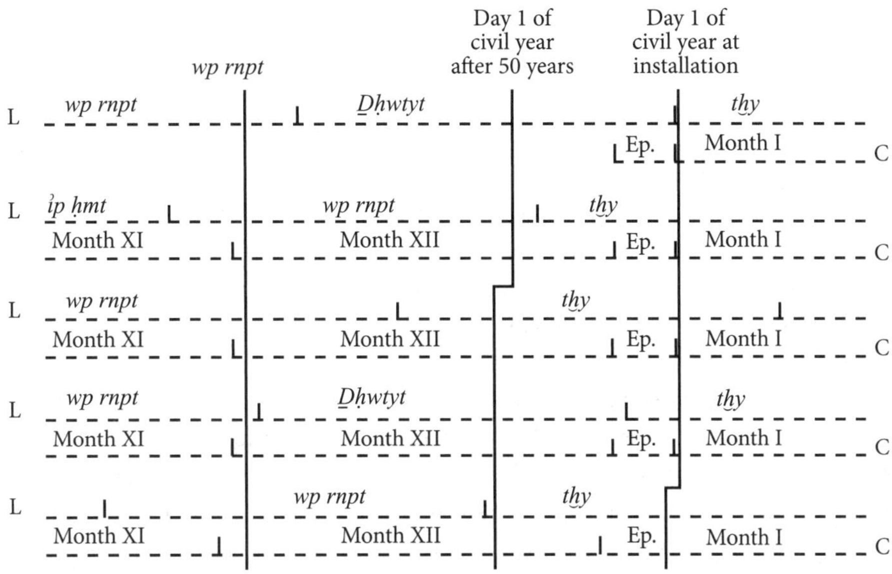


1. First, establish an intercalary month, notwithstanding that there is no direct evidence for it, either in the earlier lunar based calendar or in the later civil one.
2. Then insure that this extra month is placed between lunar month XII and lunar month I from time to time.
3. This way any lunar-based feast—say *k3-ḥr-k3*—will "bob" around its original setting "left or right," but not migrate too much back or forward in time.
4. Equally, this lunar year in which the lunar feasts operated will have a second control factor, the one established via *wp rnpt*, which Parker felt indicated the heliacal rising of Sothis.
5. And this lunar year has to be a new one. According to Parker it was the "second lunar calendar."

At this point many sympathetic readers would probably have given up on Parker's argumentum or, I suspect, accepted it without any scientific criticism. There remain, however, too many difficulties with his way of thinking, not the least of which is Parker's quandary in establishing what exactly *tpj rnpt* signified.[37]

37 See my analysis, "Calendars: Real and Ideal," in Betsy Bryan and David Lorton, eds., *Essays in Egyptology in Honor of Hans Goedicke* (San Antonio, 1994), pp. 297–308. Note 68 in Chapter 1 provides some data concerning this issue.

Outlined above in those five characteristics was Parker's "second lunar calendar." It was fixed, unlike the first, and regulated by Sothis, as Parker emphasized again and again. It was not the original lunar system in which months of varying lengths—let us say, for the sake of argument, months having twenty-nine or thirty days—occurred. Instead, it was a "refined" or "schematic" resolution to the *décalage* between the Sothis setting and the inevitable "minus eleven" days that is the lunar-solar epoch; i.e., 365 "civil" days minus 354 "lunar" days. Given an intercalary month every three or four years, the Egyptians would have "solved" their problems with the civil and lunar equations. According to Parker, this would have taken about "two centuries," and so the "second lunar calendar" would have come into existence by the Second Dynasty or so.

My analysis has reached the final stage of Parker's well-structured book. We can now see how he attempted a rationale for the names of the Egyptian civil months, many of which were based on feasts occurring on or about the first day of the civil month immediately following. He was working within a lunar-based system, of course. Hence, we can see how he got into an altercation with Gardiner over the *k3-ḥr-k3* feast and similar matters.[38] Parker had additional information that allowed him to argue for a lunar basis of this important religious ceremony. Gardiner, for his part, knew that the event was coincidental with the *nḥb-k3w* celebration—always set on V 1—but not the same. Both fought with severity over this issue, one that was rendered more complex by the overriding importance of day one of the second season, V 1, I *prt* 1, 1 Tybi.

The resolution to that conundrum was not achieved by either of the two opponents. Both knew that the key feast could take place in civil month IV as well as civil month V. Parker understood *k3-ḥr-k3* to be "flexible," but set, we now have to reiterate, within the "second lunar calendar." Hence, this auspicious celebration could "bob," as I put it earlier, move around. Indeed, it could occur before and after I *prt* 1, a fact that Gardiner also recognized but sidestepped. We can now see how the hypothesized "second lunar calendar" could resolve this situation. Because the lunar-solar epoch is 11 days, *k3-ḥr-k3* could temporarily wander, but ultimately not far, owing to the presumed intercalary month. Parker's "second lunar calendar" was self-regulating. To take the specific case in point, *k3-ḥr-k3* never would have taken place in civil month XII or month I. Nonetheless, it is now realized that Parker's evidence for any intercalary month is not extant. Therefore, we must seek another reason why the lunar festivals were not constant, and this has been offered here. This simplest explanation also accounts for the transference of the names to the civil months and has the further advantage of avoiding detailed mathematical calculations.

---

38 I am now referring to the ferocious debate between Gardiner, "The Problem of the Month-Names," pp. 9–31, and Parker, "The Problem of the Month-Names: A Reply," *RdE* 11 (1957): 85–107. Depuydt, "The Two Problems of the Month Names," *RdE* 50 (1999): 107–33, continues the discussion of the conundrum.

We can figure out the system by which a lunar event would have been transposed into the new Civil Calendar when it was introduced, bringing with it a change in month names. Let us try an example:[39]

1) A specific feast began on day one (*psḏntjw*) of lunar month V.
2) The name of that month was *K3-ḥr-k3*.
3) The name of the feast was the same, *k3-ḥr-k3*.
4) After the introduction of the Civil Calendar, the name *K3-ḥr-k3* was now applied to civil month IV.
5) But the religious event, *k3-ḥr-k3*, remained lunar-based.
6) Hence, it should have been celebrated on the first lunar day (*psḏntjw*) of month V, which had been called *K3-ḥr-k3*, but now this civil month was called Tybi.[40]
7) Therefore, *k3-ḥr-k3* would have begun to be celebrated on the first lunar day (*psḏntjw*) in civil month IV, now called *K3-ḥr-k3*.

We can now also sympathize with the Egyptians trying to determine festival dates once they had become accustomed to their civilly organized calendar. Who knew how to set an original lunar-based feast on say, *psḏntjw*, when the name of the month into which this "first day" fell was altered (by minus one)? How could they regress to an earlier stage in time? That was impossible because now everything — king, state, bureaucracy, and accountants — were subject to, led by, or organized through the Civil Calendar. Lunar-set events must have become less and less important.[41] Lunar-oriented religious celebrations may very well have become vestiges of a forgotten past.

As a lemma, perhaps the reader will now perceive the chronological difficulties that could arise, a point that bears upon the dating of the commencement of the heb sed festival.[42] In essence, to observe when lunar day one occurred, Egyptians had to

39 The following discussion works with the basic data somewhat differently than in my study "The Date of Amunhotep II's First Accession," *SAK* 40 (2010): 387–97. By an oversight in the last paper I had omitted the useful and conclusive study of Stéphane Pasquali, "La date du Papyrus BM 10056: Thoutmosis III ou Amenhotep II?," *RdE* 58 (2007): 71–86.

40 A classic Egyptological perspective on the New Kingdom names of months and feasts will be found in van Walsem, "Month-Names and Feasts at Deir el-Medîna," in Demarée and Janssen, eds., *Gleanings from Deir el Medîna*, pp. 215–44.

41 This is the thrust of my study, "The Lunar System in Festival Calendars: From the New Kingdom Onwards," *BSEG* 19 (1995): 25–40, a riposte of sorts to Alfred Grimm, *Die altägyptischen Festkalender in den Tempeln der griechisch-römischen Epoche* (Wiesbaden, 1994). This is an issue that Ptolemaic scholars need to address.

42 I can recommend a study that deserves wider readership: S. Voß, *Untersuchungen zu den Sonnenheiligtümern der 5. Dynastie: Bedeutung und Funktion eines singulären*

look at the sky. If the first day could be perceived, then all was fine with the world. *Psḏntjw* had come. Side by side with this natural event was the automatic determination of the month. Of course, the month was civil, but that is why Egyptian history had trouble with the establishment of the "proper" parameters associated with any lunar-determined event. No doubt Amunhotep III faced the same difficulty when requiring that his heb sed festival rites be properly celebrated, and we do know that they began on V 1.[43] Yet the name of that month was Tybi and not Choiak.

---

*Tempeltyps im Alten Reich* (Hamburg, Ph.D. Dissertation, 2004), pp. 124–34, but she needed to supplement her reference to Christian Leitz (*Studien zur ägyptischen Astronomie* [Wiesbaden, 1989]) with my "Calendars: Real and Ideal," (see note 37, above), "Parallelism of Thought," in C. Berger, G. Clerc, and N. Grimal, eds., *Hommages à Jean Leclant* IV (Cairo, 1994), pp. 363–77, and "Chronological Remarks," *BSEG* 11 (1998): 56. Erik Hornung and Elisabeth Staehlin, *Neue Studien zum Sedfest* (Basel, 2006), 39–40, provide the necessary background data concerning the date of I *prt* 1 (V 1) and the heb sed festival.

43 Donald Redford, *Pharaonic King-Lists, Annals and Day Books: A Contribution to the Study of the Egyptian Sense of History* (Mississauga, 1986), p. 186. Yet this scholar did not unravel the remarkable calendrical difficulty that was faced by the ruler. Cf. my "Under the Moon of Earth," in Spalinger , ed., *Revolutions in Time*, chapter 4. In general, see Murnane, "The Sed Festival: A Problem in Historical Method," *MDAIK* 37 (1981): 372 and note 22; with Hornung and Staehlin, *Neue Studien zum Sedfest*.

# 3
# The Results of the Civil Calendar

Up to now we have presented an overview of Parker's remarkable project of 1950. Except for his engagement with Gardiner, I have avoided analyzing the subsequent proliferation of research of supporting or detracting research. In hindsight, it is clear that Parker would still encounter contrary opinions aside from the monumental attack of Gardiner. This he did. One may signal out his later work on Sothic Dates,[1] the refutation of John G. Read, and his intense study of the Sothic year.[2] At this stage, however, it will be necessary to retrace our procedure and pursue two separate though intertwined aspects of Egyptian calendrics and festival dating. Fortunately, the first step has already been presented in another symposium.[3] The

---

1 The results presented in M. F. Ingham, "The Length of the Sothic Cycle," *JEA* 55 (1969): 36–40, often cited, remain imprecise. See Ronald A. Wells, "Some Astronomical Reflections on Parker's Contributions to Egyptian Chronology," in Leonard Lesko, ed., *Egyptological Studies in Honor of Richard A. Parker* (Hanover, New Hampshire and London, 1986), pp. 165–71, and Rolf Krauss, *Sothis- und Monddaten: Studien zur astronomischen und technischen Chronologie Altägyptens* (Hildesheim, 1985), passim, especially pp. 36-72. The compendium by Marshall Clagget, *Ancient Egyptian Science: A Source Book* (Philadelphia, 1989–99), though very well presented and researched, lacks Egyptological know-how.

2 I am referring to Parker's "Sothic Dates and Calendar 'Adjustment,'" *RdE* 9 (1952): 102–8, "The Beginning of the Lunar Month in Ancient Egypt," *JNES* 29 (1970): 217–20, and his final magnum opus, "The Sothic Dating of the Twelfth and Eighteenth Dynasties," in Janet H. Johnson and Edward F. Wente, eds., *Studies in Honor of George R. Hughes: January 12, 1977* (Chicago, 1976), pp. 177–84.

3 "The Beginning of the Civil Calendar," in Miroslav Bárta, Filip Coppens, and Jaromír Krejčí (eds.), *Abusir and Saqqara in the Year 2010* (Prague, 2012). This study queries the need to follow Leo Depuydt, *Civil Calendar and Lunar Calendar in Ancient Egypt* (Leuven, 1997). But see already Christian Leitz's review of that work in *BiOr* 57 (2000): 75–81. Depuydt, "What is Certain about the Origin of the Egyptian Civil Calendar?" in Hedvig Győry, ed., *Mélanges offerts à Edith Varga: "Le lotus qui sort de terre"* (Budapest,

second, on the other hand, needs greater clarification, as it appears the case that some Egyptologists remain troubled concerning the method which the ancients used to find the first lunar day.[4]

To begin, let me clear away some gorse and bracken. It will be assumed that the (heliacal) sighting of Sothis took place in the east[5] and was coordinated from the north, most probably in Heliopolis. Because scholars are generally in agreement on this matter, I need not provide the primary evidence. Yet, owing to the strictures of Rolf Krauss, it is necessary to make this statement at the beginning of the discussion.[6] It must be evident that I am not in agreement with Parker in downplaying Sothis and elevating Selene. But the very late data from the temple of Esna allow us to work out the procedure whereby the Egyptians established their Civil Calendar.

Esna explicitly places the event of *wp rnpt* in both civil month I (which was called Thoth at the time) and, in a second reference, in the month *Rᶜ-Ḥr-ȝḫtj*. The latter confused Serge Sauneron, because he saw immediately that the latter should refer to civil month XII.[7] After all, was not the "evolution," or development, of the name

2001), pp. 81–94 is both inexact and simplistic with respect to astronomical data. No "Sothic Year" could consist merely of 364 × 4 (= 1460) or 365.25 × 4 (= 1461) years. To the ancients, both hold; for us moderns, neither is valid.

4 Following the research of Pritchett, "The Athenian Lunar Month," *Classical Philology* 54 (1959): 151–57, I contributed a two-dimensional study of the matter in "Under the Moon of Earth," in Anthony J. Spalinger, ed., *Revolutions in Time: Studies in Ancient Egyptian Calendrics* (San Antonio, 1994), pp. 61–83. This will be discussed below. Krauss, "Lunar Dates," in Erik Hornung, Rolf Krauss, and David Warburton, eds., *Ancient Egyptian Chronology* (Leiden and Boston, 2006), pp. 395–431, deals with the integer problems. I will cite Leitz, *Studien zur ägyptischen Astronomie* (Wiesbaden, 1989) here but once. See Ronald Wells' review in *BiOr* 49 (1992): 723–28.

5 This is why Otto Neugebauer and Richard Parker, *Egyptian Astronomical Texts I: The Early Decans* (London, 1960), placed great emphasis upon the eastern horizon when dealing with Egyptian perceptions. The development of the original decanal system of night hours — the risings in the east of the star groups — is the prime example, and its use was based upon the eastern phenomenon of the heliacal rising of Sothis.

6 Krauss' work is cited in note 1, above. See Ulrich Luft, *Die chronologische Fixierung des ägyptischen Mittleren Reiches nach dem Tempelarchiv von Illahun* (Vienna, 1992), for example. Detlef Franke indicated that he was in support of Krauss' chronological bases in his two-part study "Zur Chronologie des Mittleren Reiches, Teil 1: Die 12. Dynastie," *Or* 57 (1988): 113–38, and "Zur Chronologie des Mittleren Reiches, Teil II: Die sogenannte 'Zweite Zwischenzeit Altägyptens,'" *Or* 57 (1988): 245–74. As it turned out, his chronology also was faulty.

7 Serge Sauneron, *Les fêtes religieuses d'Esna* (Cairo, 1962), pp. 11, 25, and 147. Add Depuydt, "Esna's Triple New Year," *JARCE* 40 (2003): 55–67. Parker, *Calendars*, p. 45, also noted some of these facts, especially the name of the last civil month in Edfu as *Rᶜ-Ḥr-ȝḫtj*.

Fig. 5 The Esna Data (reproduced from Spalinger, "The Beginning of the Civil Calendar," in M. Bárta, F. Coppens, and J. Krejčí [eds.], *Abusir and Saqqara in the Year 2010*, p. 729, fig. 3).

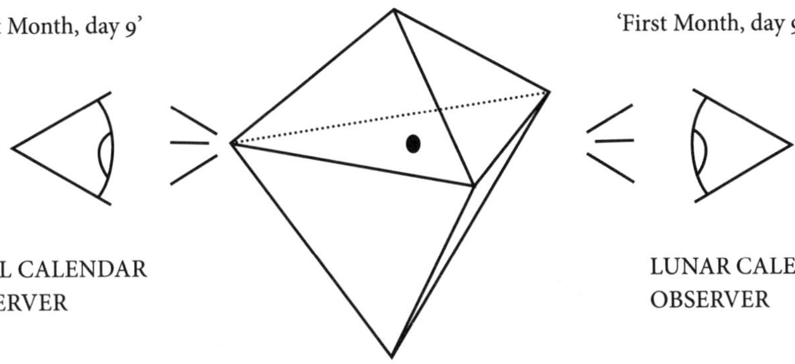


of that last civil month as follows: *Wp rnpt* > *Mswt-R*ᶜ > Mesore? (I am presenting the case in its simplest format.) The convoluted discussions of Gardiner and Leo Depuydt need not concern us here.[8] Parker attempted a rebuttal to the former, but it was not as conclusive as he wished.[9] Indeed it failed, even if most Egyptologists shunned the argument either out of respect for the two participants or through ignorance. But whatever the situation, on day 9 of *a* first month the heliacal rising of Sothis was used to determine the first day.

The additional calendrical reference that Esna supplied was that of a new *wp rnpt* set civilly on Thoth 9. The event did not indicate the name of a month but the commencement of a new year. As the festival calendar at Esna indicates, this event was known to the ancestors; it was an age-old one. Clearly, it was differentiated from the civil *wp rnpt* of I *ꜣḫt* 1. The second case, first discovered by Sauneron, located this "peculiar" *wp rnpt* in the month of *R*ᶜ*-Ḥr-ꜣḫtj*. If the latter were civil, then we would be dealing with civil month XII. Note the simple contrast: either civil month XII or civil month I. Hence, I felt that we possess, purely by chance, two key references to

8 Alan Gardiner, "Mesore as the First Month of the Egyptian Year," *ZÄS* 43 (1906): 136–44; and Depuydt, *Civil Calendar and Lunar Calendar in Ancient Egypt*.

9 The basic discussion has been referred to earlier: Gardiner, "The Problem of the Month-Names," *RdE* 10 (1955): 9–31, Parker, "The Problem of the Month-Names: A Reply," *RdE* 11 (1957): 85–107; with Depuydt, "The Two Problems of the Month Names," *RdE* 50 (1999): 107–33. No resolution of the problems incurred can be found in any of these three studies. This is due to the factors surrounding the rise of the Civil Calendar, wherein the one-month *décalage* must be considered. If so, then the difficulties seem trivial. See my "The Beginning of the Civil Calendar," in Miroslav Bárta, Filip Coppens, and Jaromír Krejčí, eds., *Abusir and Saqqara in the Year 2010* (Prague, 2012), pp. 724–35.

FIG. 6 The Calendrical Décalage (reproduced after Spalinger, "The Beginning of the Civil Calendar," in M. Bárta, F. Coppens, and J. Krejčí [eds.], *Abusir and Saqqara in the Year 2010*, p. 728, fig. 2).

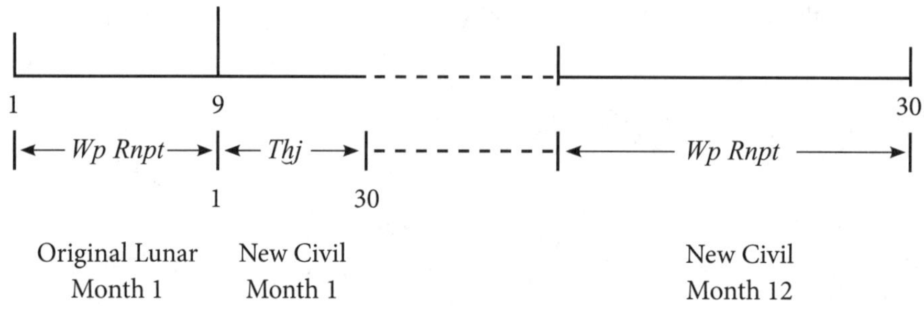


the first and same *wp rnpt* that ever existed in a civil setting. The original date was day nine of the first month, but in a calendar that was not the civil one. The latter had to have been the earlier or original lunar calendar of the Egyptians, and its first month was called *Wp rnpt*. By the time the Esna festival calendar was composed, all that was known was the barest of facts, albeit the most important of all: the Ur-*wp rnpt* was primordially set on day nine of *Rꜥ-Ḥr-ȝẖtj*, which was simply the later name for the month *Wp rnpt*, and it had to have been the first month of the then extant year. This event was also set in the first civil month, and to me this implied that it originally was located in the first month of the *same* year. Both data actually coincided.

That the famous P. Ebers insert also followed that specific date was not lost on me.[10] One could, therefore, perceive why later Egyptians, whose knowledge of the original lunar calendar was lost, had major difficulties in connecting this lunar reference, and others, within their new calendar. How could they not? But it was clear to many of us researchers that there had been a calendrical *décalage*, wherein the names of the months of the first Egyptian calendar, undoubtedly organized by the moon, were transferred over to the civil year with one difference: the month names were pulled earlier. A survey of this matter was brought forward in the previous chapter.

Now, if (lunar) day nine of lunar month one was the first *wp rnpt* of the Civil Calendar, then it would have started off with its first month being called *Tẖj*. Why not *Wp rnpt*? Because that name was already in existence and "working," so to speak, within its expected calendar, the original lunar one. True, some complication might be raised, such as the problem of the middle months, those of "Heat" (*Rkḥ wr* and *Rkḥ nḏs*), both of which too neatly appear artificial, if only because they occur in the

10 I have always referred to Parker's transcription on page 38 of his *Calendars*. It is the best known to me.

sixth and seventh civil positions.[11] Still, by linking up the Esna data with the earlier insert in P. Ebers, one can see why one Egyptian, in regnal year nine of Amunhotep I, perhaps a brilliant assistant of the king, devised a chart of twelve months commencing with *Wp rnpt*, all referring the heliacal rising of Sothis (*prt Spdt*) and (civil) day nine. The coincidence is too great to dismiss cavalierly.

P. Ebers, to the consternation of many, begins with the month *Wp rnpt* day nine and links the temporal setting with the heliacal rising of Sothis. Obviously, it appears lunar, especially as there are no overt epagomenals. (Covertly, however, they could be there.) The list of twelve months, with its specific reiteration of *prt Spdt* month by month, indicates the course of the dog star throughout its own year, the so-called "Great Year," "Sothis Year," or "Phoenix Year."[12] Therefore, the Ebers calendar is nothing more than a native Egyptian schematic survey of the star's passage through time. The insert may very well indicate that in regnal year nine of Amunhotep I, Sothis returned to its place of beginning in the cycle. But this key point concerning Esna bears upon the Ebers data to such a degree that we are compelled, whether we like it or not, to accept our new interpretation of the astounding "second" *wp rnpt* at Esna set on Thoth 9. Krauss, on the other hand, after providing new data and useful analyses, did not note the connection.[13] Once the new direction is taken, the dilemma of Buridan's Ass evaporates and the simpler solution is to be preferred, as it always is. Which is to say: on day nine of the first lunar month, then called *Wp rnpt*, and in the original lunar calendar of course, the first civilly important *prt Spdt* took place. This was the date of the inauguration of the Civil Calendar, and it was remembered over centuries, as Ebers and Esna prove.

Now we can turn to the determination of lunar events during the period when Egypt had a Civil Calendar. This is as straightforward as the above argumentation was. Once more, complex interweaving of time's warp and woof is not necessary. Because the Civil Calendar regulated life, it was paramount. Yet, both we and the ancients would always like to know: can one predict the first (Egyptian) lunar day?

---

11 Parker, *Calendars*, p. 46 with note 100. The name of the eighth civil month, *Rnwtt*, "Harvest," should have alerted one to the fact that the designation should have heralded or commenced a season. Once the *décalage* of "one month minus" is understood, then it is seen that this name occupied its proper position within the earlier Egyptian lunar year. Cf. Parker, ibid., p. 45, and Spalinger, "Notes on Ancient Egyptian Calendars," *Or* 64 (1995): 17–32.

12 The continual "standard" year of 365 days — an Egyptian civil year, of course — in which the phoenix lives, is to be expected. There are interesting comments on this factor in Jean Hubaux and Maxime Leroy, *Le mythe du phénix dans les littératures grecque et latine* (Paris, 1939), pp. 8–9.

13 *Sothis- und Monddaten*, pp. 104–16. One can see the same conundrum reflected in a series of articles on the Ebers insert by Krauss, Leitz, and Jürgen von Beckerath in *ÄuL* 3 (1992).

That is extremely hard; witness the rise of the Metonic Cycle and, of course, the schematic data from P. Carlsberg IX.[14] But no predictive calendrical model is known to us from pharaonic Egypt until very late, and for our interests we can exclude the time period preceding the Persian conquest of Egypt in 525 BC.[15] Hence, the issue is simple. If no absence of a lunar crescent in the morning[16] is recorded on day 28 of a lunar month, then there must be at least twenty-nine days in that month. But one cannot state that the given lunar month has only twenty-nine days. If, in the morning of day thirty (by counting) no crescent is seen, then the month had twenty-nine days; if not, the lunar month had thirty days. This is what Bradley Schaefer and, independently, Ronald Wells have stressed. Namely, that there is no way of predicting when a month is hollow or not. The same thing would have occurred in Athens — *mutatis mutandis*[17] — until some type of artificial system was developed, a point well-established by William Pritchett.[18]

---

14 Alan C. Bowen and Bernard R. Goldstein, "Meton of Athens and Astronomy in the Late Fifth Century B.C.," in Erle Leichty, Maria deJ. Ellis, and Pamela Gerardi, eds., *A Scientific Humanist: Studies in Honor of Abraham Sachs* (Philadelphia, 1988), pp. 39–81. The Babylonian influence, proven by the two authors, is incontestable. For additional background information on this issue, see notes 14–15 of Chapter 2.

15 I have chosen this date on the possibility that a Babylonian system of a twenty-five year cycle entered Egypt during the Achaemenid Period; i.e., preceding the evidence of P. Carlsberg IX. See my remarks in "Calendrical Comments," *BiOr* 51 (1994): 5–20, and more recently Chris Bennett, "Egyptian Lunar Dates and Temple Service Months," *BiOr* 65 (2008): 525–54.

16 The morning epoch was also used in Coptic (as expected) and Ethiopic (following the Alexandrian norm) time reckoning: Pritchett, "The Calendar of the Gibbous Moon," *ZPE* 49 (1982): 263, note 61. See as well Grover Hudson, "Ethiopian Month-names," in Gábor Takács, ed., *Semito-Hamitic Festschrift for A.B. Dolgopolsky and H. Jungraithmayr* (Berlin, 2008), pp. 150–54.

17 In particular, the remarks contained in William Pritchett and Otto Neugebauer, *The Calendars of Athens* (Cambridge MA, 1947), are important to read. The study has not become outdated in that respect. Note once more the affiliation to Neugebauer. Pritchett was studying at Princeton when Neugebauer contacted him. (Neugebauer was always associated with that University.)

18 His basic article is referred to above in note 4; add his "Calendars of Athens Again," *BCH* 81 (1957): 269–301, "The Calendar of the Athenian Civic Administration," *Phoenix* 30 (1976): 337–56, and "Postscript: The Athenian Calendar," *ZPE* 128 (1999): 79–93. It has to be noted in passing that Pritchett and Neugebauer opposed many of the Classical scholars — who were non-specialists in these astronomical matters, of course — who maintained that the Athenians used a Metonic system or cycle. See Pritchett's nearly last piece of scholarship, "The Solar Year of Thucydides," in his *Thucydides' Pentekontaetia and Other Essays* (Amsterdam, 1995), pp. 173–204, and especially pp. 193–94. He referred to the important study of Goldstein and Bowen, "The

Interestingly, Neugebauer and Parker refer to the appearance of the moon in Egypt when the second day (*ꜣbd*) had come.[19] On that Egyptian day and for the first time in an Egyptian lunar month, the visible moon "preceded all the other appearances throughout the month."[20] Neugebauer's Babylonian data had the moon visible during twenty-eight nights in eighty percent of the tabulated months. The remaining twenty percent were covered by lunar visibility with a month composed of either twenty-seven or twenty-nine nights.[21] But the sources that Neugebauer depended upon were not naked eye sightings — they are not astronomical observations. Rather, they are ephemerides, computed tables of predicted lunar and planetary positions. In fact, the sequences of twenty-seven, twenty-eight, and twenty-nine visible appearances of the moon per lunar month are arithmetical predictions.

In his "Moonwatch" experiment, Bradley Schaefer used the Babylonian system and showed that it was not reliable for computing predicted visibilities.[22] Schaefer was appropriately subdued in his remarks concerning the unfortunate fact that no set of empirical month lengths can be matched with or by calculations. He added that there can easily be multiple agreements through a range of, say, most of a century. There is, at best, a low statistical significance between the matches in any set of twelve Egyptian month lengths by calculations and by observation.[23] Most often Schaefer noted the thin crescent visibility for observations, but I think that the situation is even worse

---

Introduction of Dated Observations and Precise Measurement in Greek Astronomy," *Archive for History of Exact Sciences* 43 (1991): 93–132. Add their other joint work, "On Early Hellenistic Astronomy," *Centaurus* 32 (1989): 272–93.

19 Neugebauer and Parker, *Egyptian Astronomical Texts* I, p. 80, with note to VII 24–25.

20 Ibid.

21 Data taken from Neugebauer, *Astronomical Cuneiform Texts* III (New York, Heidelberg, and Berlin, 1983), pp. 140–50. Cf. John M. Steele, "The Length of the Month in Mesopotamian Calendars of the First Millennium BC," in John M. Steele, ed., *Calendars and Years: Astronomy and Time in the Ancient Near East* (Oxford, 2007), pp. 133–48. Add Paul-Alain Beaulieu, "The Impact of Month-lengths on the Neo-Babylonian Cultic Calendar," *ZA* 83 (1993): 66–87.

22 See Ronald Wells, "The Role of Astronomical Techniques in Ancient Egyptian Chronology: The Use of Lunar Month Lengths in Absolute Dating," in J. M. Steele and A. Imhausen, eds., *Under One Sky: Astronomy and Mathematics in the Ancient Near East* (Münster, 2001), pp. 459–71.

23 In particular, see Bradley Schaefer, "Lunar Crescent Visibility," *Quarterly Journal of the Royal Astronomical Society* 37 (1996): 759–68, and "The Heliacal Rise of Sirius and Ancient Egyptian Chronology," *Journal for the History of Astronomy* 31 (2000): 149–55. The interesting statistical analysis of Sebastian Porceddu, Lauri Jetsu, Tapio Merkkanen, and Jaana Toivari-Viitala, "Evidence of Periodicity of Ancient Egyptian Calendars of Lucky and Unlucky Days," *CAJ* 18 (2008): 327–39, deserves a detailed analysis that I cannot present here.

when one tries to observe *psḏntjw*, the "null case." This problem is compounded by the fact that the number of days between the last crescent visibility and new crescent visibility varies, a point that Edward Wente never considered.

There being no known way that the Egyptians could have approximated the day called *psḏntjw*, how did they operate? I presume at this juncture that it is obvious the numbering of any lunar day depended upon the establishment of the first. Hence, we can narrow the discussion to *psḏntjw* alone. Because we have eliminated the possible existence of a "second lunar calendar," although it is one that Krauss still follows,[24] we can have recourse only to the naked eye, the limitations of which Schaefer and Wells have indicated.[25] In a previous study over fifteen years ago, I attempted to stress the remarkably unimportant position that moon-oriented events had in the records of the Ptolemaic and Roman temples.[26] It is not surprising to me that lunar-based feasts had virtually disappeared from the official temple calendars as late as, say, the time of the early Ptolemies. I now feel that the presentation was too conservative, and it is in need of no small degree of refinement.

Certainly by the New Kingdom — I am sidestepping the evidence from Dynasties XI–XIII — the number of annual religious celebrations regulated by the moon was limited. In the lengthy festival calendar of Medinet Habu, for example, one can refer to the Valley Festival and only two others.[27] But an excellent case in point, which I shall employ as a paradigm, is the ever-consulted reference of Thutmose III on his black granite stela dated to regnal year twenty-four.[28] The event being commemorated was the beginning of the construction of the king's festival temple at Karnak.[29] And it was Edward Wente who upset previous chronographers and students of absolute chronology and postulated a new interpretation to a key section in the inscription.[30]

24 *Sothis- und Monddaten*, pp. 15–19.

25 See the material cited in note 11 to Chapter 2.

26 "The Lunar System in Festival Calendars: From the New Kingdom Onwards," *BSEG* 19 (1995): 25–40.

27 Spalinger, "Chronological Remarks," *BSEG* 22 (1998): 56–57. For the lunar orientation, see *KRI* V 123.5 and 180.9 with the Min festival: Parker, *BiOr* 9 (1952): 102; Gardiner, review of J. G. Frazer, *The Golden Bough: Adonis, Attis, Osiris. Studies in the History of Oriental Religion* (London, 1914), *JEA* 2 (1915): 125 with note 3; Henri Gauthier, *Les fêtes du dieu Min* (Cairo, 1931), passim, but especially chapter 2; Parker, *Calendars*, pp. 47–48; Krauss, *Sothis- und Monddaten*, pp. 142–43; and *KRI* V 182.4 with 201.2.

28 *Urk.* IV 833–38; cf. Jürgen von Beckerath, "Ein Wunder des Amun bei der Tempelgründung in Karnak," *MDAIK* 37 (1981): 41–49; add William Murnane, "The Sed Festival: A Problem in Historical Method," *MDAIK* 37 (1981): 373, and note 29.

29 See Jean-François Carlotti, *L'Akh-menou de Thoutmosis III à Karnak: étude architecturale* (Paris, 2001).

30 Edward Wente, "Thutmose III's Accession and the Beginning of the New Kingdom," *JNES* 34 (1975): 265-72.

His argument was purely philological, and it depended upon the use of Second Tenses as developed by Hans J. Polotsky since 1943.[31] From his determination that the adverbial adjunct in the key sentence containing a date was to be emphasized, Wente concluded that the given "formal date" was not that of the new moon (*psḏntjw*) in regnal year 24 of Thutmose III but, instead, the preceding day. Hence, its new chronological importance meant that a revision of earlier new moon calculations for this pharaoh was necessary.

It should come as no surprise that, at least in the New World, it was at the University of Chicago that Polotsky's work was first received and approved.[32] One might then surmise that Wente's analysis stands or falls upon philology. It does not. Rather, as Peter der Manuelian saw, Wente's assumptions were too broad for his conclusions.[33] Perhaps others should have followed this scholar's early foray in Egyptology.

The issue is simple. Wente implicitly believed that the Egyptians could tell when a new lunar first day, *psḏntjw*, could occur. According to Wente, the inscription indicated that Thutmose III was awaiting (*ḥr sꜣwt*) the lunar first day. The implication is that *psḏntjw* was to come, but when? In line seven, it is added that the temporal setting was on the last day of the sixth month, a portentous date, since it is exactly in the middle of the twelve Egyptian civil months. Surely, that factor needed to be discussed in any philological and lexical commentary. Thus Wente should have realized, although he did not mention this in his study, that on II *prt* 30, *psḏntjw* was in abeyance by just one day. There are only two possibilities open to us:

1) Either day 30 of civil month six, Mechir, was the then known last day of a lunar month, or
2) It was not.

Let the first day of civil month VII, Phamenoth, coincide with the first lunar day of the following lunar month. That being so, what is the probability that a particular day of a civil month = the same numbered lunar day? For the sake of argument let us exclude the epagomenals and work with a 30/30 combination. Then the odds are 1/30 or 3 1/3 percent, an extremely low figure.[34] Always one must be careful of

31 Classically, Hans Jacob Polotsky, *Études de syntax copte* (Cairo, 1944).

32 Notable early use of Polotsky's Second Tense analysis by the Egyptological Chicago students and scholars include the following two studies: Ronald James Williams, *Morphology and Syntax of Papyrus Insinger* (Chicago, Ph.D. Dissertation, 1948); and Parker, "The Function of Imperfective *sdm.f* in Middle Egyptian," *RdE* 10 (1955): 49–59.

33 Peter der Manuelian, *Studies in the Reign of Amenophis II* (Hildesheim, 1987), pp. 15–19. The study was completed while the author was working at Chicago House, Luxor.

34 The "real" result is more complicated. Krauss, *Sothis- und Monddaten*, p. 32, presents an analysis that needs some refining: "The probability that n different civil dates in-

simple statistical analysis.[35] Should we accept the tiny mathematical ratio? Note that my argument does not reside upon Wente's reading of the key passage: "It was in Year 24, II *prt* 30 (being) the day of the tenth day feast of Amun in Karnak,[36] that

---

dependent of one another coincide with the Xth day of any lunar month is defined by the product of the individual probabilities. If the coincidence of every single date with each lunar month day is considered as equally likely, then the probability amounts to $(1/30)^n$." This is correct. He continues: "As an example, for n = 3 the probability that three sequentially independent civil dates fall on a first lunar day is $(1/30)^3 = 27000$." He means 1/27000 or $3.7 \times 10^{-5}$. Krauss continues: "This corresponds to the probability—of small expectancy—of throwing the same number [on a die: my comment] 5–6 times in an uninterrupted row. Other probabilities for the coincidence of civil dates with dates of the lunar calendar follow if it is known that civil dates and lunar month days do not fall accidentally together but rather a casual connection exists between both date groups. One of these types of causalities can exist because in ancient Egypt dates (i.e., cult acts) were fixed from selected lunar month days. As is well known the 1st, 2nd, 6th, 7th, 15th, and 23rd lunar month days came into question in the New Kingdom, but the order of these cult dates is uncertain. The probability that a traditional civil date coincides with one of these 6 lunar days is w = 1/6." This depends upon how much a civil month and lunar month are out of synch with each other, but since the one moves against the other so that each day in turn of the lunar month can coincide with a particular civil date, then 6/30 = 1/5 is perfectly valid. This is identical to the statistics of drawing one red ball on the first try from a bag of thirty balls, six of which are red. Wells put it another way in "The Role of Astronomical Techniques in Ancient Egyptian Chronology: The Use of Lunar Month Lengths in Absolute Dating," in John M. Steele and Annette Imhausen, eds., *Under One Sky: Astronomy and Mathematics in the Ancient Near East* (Munster, 2002), p. 461: "This procedure ignores a basic mathematical point, namely, that the matching of any set of Egyptian month lengths in any century is of low statistical significance because there will always be multiple matches throughout the range. So even assuming perfect ancient recording of lunar dates, month sequences from multiple documents written in the same era like those from Illahun cannot be assigned the appropriate fixed yearly interval solely on the basis of computed sequences. Moreover, a statistically valid determination of an absolute date by this method, assuming perfectly observed, error-free data, would require between 250 and 300 precisely spaced month lengths according to Huber who has examined this procedure for Babylonian cuneiform texts!" The reference to Peter J. Huber is: "Modeling the Length of Day and Extrapolating the Rotation of the Earth," in Fabrizio Bònili, Salvo de Meis, and Antonio Panaino, eds., *Astronomical Amusements: Papers in Honor of Jean Meeus* (Milan, 2000), pp. 91–104. One final point: in footnote 1 on page 32, Krauss refers to Maurice Alliot, *Le culte d'Horus à Edfou au temps des Ptolémées* I (Cairo, 1954), pp. 176–79, with regard to these six presumed lunar days. Krauss worked with the "Ritual of Amunhotep I." Why could not Medinet Habu's *ḥbw nw pt* have been employed as well? There are eight—instead of six—of these days there present: 29, 30, 1, 2, 4, 6, 10, and 15.

35 Follow-up statistics may be read in Krauss, "Die Mahler-Borchardtsche These über die Apis-Inthronisationen bei Vollmond," *Acta Praehistorica et Archaeologica* 39 (2007): 339–48. His point to prove: that five known dates of Apis enthronements occurred within ± 3 days of the full moon; i.e., during a 7-day stretch centered on the full moon. Yet binominal statistics work only for random distributions, and the 7-day stretch is not random. It occurs precisely in the center of the month, not at one end or the other, nor as 7 randomly distributed days through the 30. They are immutably linked together. Krauss' interesting analogy only applies if the 7 white balls can be randomly distributed through the remaining 23 black ones. But they cannot. A good parallel: give someone a 30-chamber revolver in which the first 11 chambers contain bullets, the next 7 are blanks — i.e., empty chambers — and the remaining chambers are also loaded with bullets. Spin the chamber. Then let the person point the gun at his head and pull the trigger. What is the probability that he will kill himself? Far greater than were the 7 empty chambers originally scattered randomly. Check with http://stattrek.com/Lesson2/Binomial.aspx. Using the calculator, put in 5 trials — 0.766666 probability of 1 success — and let the number of successes range from 0,1,2,3,4,5, and you reproduce Krauss' table 9. However, the key is "random." You can only compute the probability of success of 1 trial if the 7 white balls are randomly distributed, and they are not. So you cannot use the figure of 7/30.

36 One is automatically reminded of Hatshepsut's "oracle" for kingship: see Jean Yoyotte, "La date supposée du couronnement d'Hatchepsout," *Kêmi* 18 (1968): 85–91, and Pierre Lacau and Henrie Chevrier, *Une chapelle d'Hatshepsout à Karnak* (Cairo, 1977), pp. 133–35. The passage is as auspicious as, if not more so than, the Thutmose III case. Others could be added, but I wish merely to highlight the connection between key calendrical dates and events to come. The text reads: "Regnal year II, second month of Emergence (*prt*), day 29, the third feast of Amun — corresponding to the Litanies of Sechmet, the second day — The Two Lands were foretold (*sr*) for me…." Additional material can be assembled at this point, but let me include only the following pertinent data associated with the Opet festival and oracles: Pascal Vernus, "Un texte oraculaire de Ramsès VI," *BIFAO* 73 (1975): 103–10 and *KRI* VI 282.14–83.11; with Amin A. M. A. Amer, "Reflections on the Reign of Ramesses VI," *JEA* 71 (1985): 66–70; Charles F. Nims, "An Oracle in the 'Repeating of Births,'" *JNES* 7 (1948): 157–62 with *KRI* VI 702.5–703.3 — an excellent example; von Beckerath, "Die 'Stele der Verbannten' im Museum des Louvre," *RdE* 20 (1968): 7–34, with now Karl Jansen-Winkeln, *Inschriften der Spätzeit. Teil I: Die 21. Dynastie* (Wiesbaden, 2007), pp. 72–74; and the pertinent remarks in Jean-Marie Kruchten, *Le grand texte oraculaire de Déhoutymose intendant du domain d'Amon sous le pontificat de Pinedjem II* (Brussels, 1986), pp. 23–26, 57–60, and 246–64. Chapter 5 presents a detailed yet handy analysis of the chronological ordering of the main text. Add pp. 242–44 in his *Les annales des prêtres de Karnak (XXI$^{e}$–XXIII$^{e}$ dynasties) et autres textes contemporains relatifs à l'initiation des prêtres d'Amon* (Leuven, 1989). See the important year three text of Ramesses II: *KRI* II 345.11–47.4 and Mahmud Abd el-Razik, "The Dedicatory and Building Texts of Ramesses II in Luxor Temple," *JEA* 61 (1975): 126, note 18, for yet another auspicious *psḏntjw*. Yet is the date of the commencement IV 30? See *KRI* II 347.2 with note b — "tr." with the restored 10

my Majesty gave the order to prepare for the stretching of the cord while awaiting the day of the new moon in order to stretch the cord for this monument."[37] Instead, it focuses upon the date within the civil year, the key technical word *psḏntjw*, and argues from the significant timing of the event; namely, the first day of the seventh month. A second parallel auspicious coincidence occurred in the plastered addition to the Medinet Habu Calendar. The official celebration of remembrance of the Meshwesh defeat in the eleventh regnal year of Ramesses III was set on the first day of the second season, the "second *wp rnpt*," I *prt* 1.[38]

In order that Wente's chronological position be saved, and this includes Parker's earlier analysis as well, one simply advances from day 30 in civil month VI to the first day in the following month. Moreover, the latter presents civil day 1 = lunar day 1. Is this second position valid? The probability of any date being lunar or civil is 1/2, 50 percent. That is to say, independent of the coincidence of integers, a day is either lunar or civil. *Tertium non datur*. But is it correct that the name of an Egyptian day, and not a number, proves that the day under consideration has been determined by the moon's cycle? When we say or write, for example, February the seventeenth, February seventeen, the seventeenth of February, or seventeen February, we are not abbreviating with numbers. Yet with 17/2, 2/17, February 17th, and so forth, we are providing a faster method of communication. While there is a difference between our manner of presenting dates and the Egyptian system of differentiation — say between *psḏntjw* and 1, or *prt sm* and 4 — I do not believe that this distinction affects our argument.[39] The *names* of the Egyptian days, all of which are *assumed* to refer to lunar events, indicate their dependence upon Selene; Parker's discussion is worthwhile to review.[40] See, for example, his coverage of the clear-cut cases of the first and last quarters of the moon month, the "arrival days," and so forth.

However, if it were the case that any of these names were used within a civil context, then the carefully built up and consistently maintained edifice of lunar dating would

of el-Razik "very dubious." Actually, *ʿrqj* seems best! El-Razik postulated "Day 10(?) + x" on page 126 with note 25. This, however, appears impossible. One reason why I opt for IV *ꜣḫt* 30 is that the following day commences the second season and we have civil day 1 = *psḏntjw*. Krauss briefly covers the Opet feast and dating on p. 152 of his *Sothis- und Monddaten*.

37 "Thutmose III's Accession and the Beginning of the New Kingdom," 266.

38 Spalinger, "Chronological Remarks," 56. This was not understood by Haring, *Divine Households: Administrative and Economic Aspects of the New Kingdom Royal Memorial Temples in Western Thebes* (Leiden, 1997), pp. 70-1.

39 Once more, Parker's "eclectic" (his word) List 2 in his *Calendars*, p. 11-12, more than suffices for this aspect of the discussion.

40 Ibid., 12–13; add Spalinger, *The Private Feast Lists in Ancient Egypt* (Wiesbaden, 1996), p. 171, note 6.

be adversely affected.[41] Why, for example, could not a simple passage such as "regnal year three, second month of inundation, day *sjꜣw*" indicate what, for Egyptologists or ancient Egyptians, is in shorthand, "regnal year 3, month 2 of Inundation, day 14?" Consider the still useful remarks of Gardiner with respect to civil dating.[42] In one example he transliterates a dated passage of Amenemhet III as "*ḥꜣt sp 2 (ꜣbd) 3(-nw n) ꜣḫt sw 1*." Note in particular his additions between two sets of parentheses. Whereas Gardiner was opposed to translating *(ꜣbd) 4(-nw n) prt sw 5* as "the 5th of Pharmouthi," to quote him,[43] he observed that in such datings human experience allowed for or permitted abbreviations. Surely, the same can be said regarding the integer writings for the days themselves. In fact, if it can be shown that a *name* for a civil day could be written in place of a *number*, then the issue of Thutmose III's presumed lunar date has to be *sub judice* at best. Is this not so?

If the case can be proven that a name of the day could have been rendered for a number, then all named days must be examined with particular care. I would not worry about hieratic accounts, because their work-a-day annotations are the mental and physical conundrums and solutions that the Egyptian scribes encountered, defeated, or resolved.[44] Monumental texts, especially royal ones, are more suspect, I feel, although even there it is the context that determines the outcome. Yet the existence of coincidences cannot be swept under the carpet, and this is what we have to face when dealing with the presumed Thutmose III lunar day of regnal year 24.

Fortunately or otherwise, there is a counter-example, first brought to attention by Heinrich Brugsch in 1872.[45] Parker embedded the reference in his footnotes, where few have seen the implications.[46] Once a possibility is admitted, as the proverbial camel's nose or whatever, we are forced to entertain its existence elsewhere. In this

---

41 Krauss depended upon Wente in his *Sothis- und Monddaten*, pp. 121–23; see now his subsequent "Lunar Dates," in Hornung, Krauss, and Warburton, eds., *Ancient Egyptian Chronology*, pp. 420–22.

42 *Egyptian Grammar*³ (Oxford, 1957), pp. 203–6 (Excursus C).

43 Ibid., p. 205.

44 Thus, in "Praise God and Pay the Priests," *ÄuL* 7 (1996): 43–57, I discussed the accounts from one Illahun document in an effort to refute the basis of lunar datings. I remain convinced that Joseph Cavour's enemies — see H. G. Wells — were much too powerful for many Egyptologists to disregard.

45 Heinrich Brugsch, "Bau und Maaßse des Temples von Edfu," *ZÄS* 10 (1872): 14, republished in *Edfou* IV 11.4. In conjunction with this datum and the remarks of Brugsch, add the helpful references in *Edfou* IV 2.10, VII 6.3–4 and 8.7–8; Constant de Wit, "Inscriptions dédicatoires du temple d'Edfou," *CdE* 36 (1961), 56–97 and 277-320; and the very useful study of Sylvie Cauville and Didier Devauchelle, "Le temple d'Edfou: étapes de construction. Nouvelles données historiques," *RdE* 35 (1984): 31–55.

46 *Calendars*, p. 71, note 30.

case there is a reference wherein a civil day was named. Moreover, mere acceptance of a 50-50 possibility of lunar versus civil reduces the probabilities listed earlier by a multiplicative factor of 1/2, thereby rendering the assumption of a lunar date nugatory. It is up to chronologists, then, to argue with better force elsewhere.

On the other hand, it is necessary to avoid extraordinary interpretations that can be made when one assumes that a *name* of a month, which is embedded within a text that is civil in orientation, indicates that a lunar calendar was employed. After the introduction of the Civil Calendar in the early third millennium BC, it was no longer necessary or desirable to use the increasingly archaic moon-based system of calendrics outside of specific cultic or religious settings. Harry James was correct to view the key three *names* of months that are to be found in the Hekanakhte Papyri as civil.[47] What else could they be? The writer of these letters was dealing with a series of events outside of any temple complex. Hence, he employed the Civil Calendar. As we do, Hekanakhte could write (or speak!) the specific name of a month — e.g., our "June" or "July" for example — instead of 05 or 06, 5 or 6, etc. I do not feel that James Allen's interpretation should be followed.[48] Indeed, it is significant that in those

47 T. G. H. James, *The Ḥekanakhte Papers and Other Early Middle Kingdom Documents* (New York, 1962), pp. 3–4.

48 James Allen, *The Heqanakht Papyri* (New York, 2002), pp. 135–36. He provides no proof for his surprising position except to refer to Depuydt's assertions in *Civil Calendar and Lunar Calendar in Ancient Egypt*. Unfortunately, on page 257, note 5, the situation is made even more confusing by his attempt to determine the length of lunar months by means of a "commercially available program (SkyMap)." This is scientifically unacceptable. Probably the simplest and best website to use for historians is that by Fred Espanak, who is a specialist in eclipses. Indeed, his six millennia of lunar phases ought to be a "godsend" for Egyptologists, none of whom use the site as far as I know. Its earliest year however is -1999, or 2000 BC: http://eclipse.gsfc.nasa.gov/phase/phasecat.html. Espanak's work was dependent upon Jean Meeus' *Astronomical Algorithms*[2] (Richmond VA, 1999), which is the best compilation of the equations necessary for the computations. Note also that Espanak gives the values for Delta-T. See Wells, "The Role of Astronomical Techniques in Ancient Egyptian Chronology: The Use of Lunar Month Lengths in Absolute Dating," for a detailed analysis of crucial observational points that prevent any certainty being obtained. He pointed out and explained why last and first crescent sightings are most difficult to observe and, hence, to determine for ancient times. Particularly telling is the comparison Wells made in his fig. 3 between the modern "Moonwatch" observations of first crescents with the theoretical predictions of Fotheringham, Maunder, and Illyas based on Greek observations of more than a century ago. The comparison shows that reality and theory are not even close. Chris Marriot, who wrote the SkyMap planetarium program, provides the following details on the algorithms used to compute the lunar and planetary ephemerides as found in "Data Sources" of the Skymap Pro "help" section:
a) "The position of the Moon is computed from the ELP 2000-82B lunar theory (M.

Chapront-Touzé and J. Chapront, 1988). The full theory contains some 37,000 terms; we ignore all terms smaller than 0.001", thus reducing the number of terms which have to be computed to about 10,000. The resulting position of the Moon agrees with that computed using the full theory to within 5–10 metres, which is more than adequate for our needs. Spot checks against lunar positions tabulated in the 'Astronomical Almanac' give agreement to better than 0.01."

b) "Because of the irregularities in the Earth's rotation, the theories of motion of astronomical bodies do not use Universal Time (GMT), but a uniform timescale called Terrestrial Dynamical Time (TDT). SkyMap uses TDT internally for all its calculations of planetary positions, etc., but obviously the user specifies the time for which a map is required in UTC (or rather, in local time, which is converted to UTC). The difference between TDT and UT is called 'delta T,' and currently has a value of approximately 1 minute. It is currently increasing at a rate of somewhat less than 1 second per year. The problem is that the value of delta T can only be determined historically (typically by analyzing the motion of the Moon), and current and future values can only be estimated, whilst values for the distant past (before the advent of modern astronomy) are uncertain to the order of many minutes. What this means in practice is that although the time of a total solar eclipse in the year 1500 BC could be computed to a precision of a fraction of a second in TDT, the actual time in UT that the eclipse occurs (hence the places on Earth from which it is visible) will be uncertain to within several minutes. The 'Astronomical Almanac' lists the values of delta T for every year from 1620 onwards (currently up to 1996), and provides estimates of its value for the current time. SkyMap has all this data stored, and interpolates or extrapolates in this table to find values of delta T for dates between 1620 and 2002."

c) "For dates prior to 1620, an estimate is made using the method of F. R. Stephenson and M. A. Houlden, *Atlas of Historical Eclipse Maps*, Cambridge University Press (1986). They estimate the uncertainty to be 15 minutes at 1500 BC."

However, Allen's lunar dates are nevertheless unreliable. For general historical, archaeoastronomical work the program is fine. But as to the computation of first and last crescents, Wells' comments imposed by the observational uncertainties still hold true. Skymap will give a reasonably accurate answer for the date of New Moon, or any other phase (i.e., First Quarter, Full Moon, Last Quarter), but *not* for the appearances of the last and first crescents. The latter still depend on the vagaries that Wells listed, and since the time interval between the two is highly variable (3, sometimes 4, more rarely 5 days) as Parker (*Calendars*) himself long ago noted, then the determination of first crescent or last crescent will be as uncertain as he described therein. One must remember that *psḏntjw* is not the same as "new moon." Parker-defined *psḏntjw* — the beginning of the ancient Egyptian lunar month — as the morning on which the last crescent moon could no longer be seen in the eastern sky before sunrise. This would nominally be the day after "last crescent," but since the determination of the morning of last crescent is itself uncertain, then so will be the start of the lunar month.

papyri the first day of a named month is written by *wp* and not *psḏntjw*.[49] When lunar day one is indicated, then *psḏntjw* is always written. *Wp*, on the other hand, never occurs in a clear-cut reference for the first lunar day.[50]

A further intriguing and perplexing case of lunar and civil calendrics, also surveyed by Parker, is connected to the previous discussion. I am referring to the difficulty that he had in making his case to Gardiner.[51] First, let me survey his statements in *Calendars*.[52] There he set up an equation between the fixed (civilly speaking) celebration of *Nḥb-k3w* and the well-known *k3-ḥr-k3* feast. Subsequently, he wrote a manuscript for publication in the *Journal of Egyptian Archaeology*, and it was permitted to be sent to Gardiner. Parker's understanding of the matter of I *prt* 1 circled, once more, around the lunar orientation of *k3-ḥr-k3*. But that time, a few years after the publication of *Calendars*, he turned to the evidence from the Donation Stela of Ahmose-Nofretari.[53] The date of that monumental inscription, set up in the forecourt of the temple of Amun but later reused as fill, was day 7 in the fourth month, civil *K3-ḥr-k3*. The text is now available in a standard edition of Wolfgang Helck, although one should always follow the excellent facsimile of Gerhard Haeney.[54] Betsy Bryan proposed that Ahmose-Nofretari bought the office of Second High Priest of Amun, although this is by no means certain.[55] This explanation might fit the purpose of the stela, but there is an outstanding lacuna that prevents any conclusive proof.[56] Helck in fact argued for a position opposite to Bryan's, and this conclusion is generally adhered to.[57] For our interests the two passages of significance are:

---

49 Henry Fischer, *Varia Nova (Egyptian Studies III)* (New York, 1996), pp. 191, 230–31, discusses this *wp*. One obvious point: if Hekanakhte used a lunar month, how was it determined? By him alone? Who followed him? By him and friends? Or by a temple, and, if so, which one? Once more, we can see with clarity that the lunar hypothesis is impossible. By the Middle Kingdom, Egypt was no "archaic state."

50 This is easily seen on plate 48 in Allen's edition (Account VII [MMA 22.3.522], recto 11): *m wp n Šf-bdt*, and on plate 30, Letter II (MMA 22.3.5127), recto 32: *m wp n ḫntj-ḫtj-prtj*.

51 The basic sources are presented in note 8, above.

52 Pages 41 and 58 provide the best summary.

53 I always prefer the excellent facsimile by Gerhard Haeney in Ibrahim Harari, "Nature de la stèle de donation de fonction du roi Ahmôsis à la reine Ahmès-Nefertari," *ASAE* 56 (1959), pl. II.

54 Wolfgang Helck, *Historisch-biographische Texte der 2. Zwischenzeit und neue Texte der 18. Dynastie* (Wiesbaden, 1975), pp. 100–103.

55 "Property and God's Wives of Amun," at a conference called "Women and Property," organized by D. Lyons and R. Westbrook at Harvard University, 2005.

56 "Überlegungen zur Geschichte der 18. Dynastie," *OrAnt* 8 (1969): 286.

57 Cf. Michel Gitton, *L'épose du dieu Ahmès Néfertary: documents sur sa vie et son culte posthume* (Paris, 1981), pp. 7–11, and *Les divines épouses de la 18$^{e}$ Dynastie* (Paris, 1984),

a. ([Regnal year X], fourth month of Inundation, day 7 ..... (column 1)
b. ... during the festival of ⸢*k3-ḥr*⸣-*k3* .....(column 21).

Parker, who now felt that the event lasted seven days, at least during the New Kingdom, combined the Ahmose-Nofretari data with a new piece of evidence.[58] This was a literary ostracon from Deir el Medineh, first published by Georges Posener.[59] Parker left open two possibilities of chronological interpretation. In the first case, he placed the event in the interval of days 1 to 7 of the fourth civil month. In the second, he stated, "It is just as possible, of course, that it began on IIII *3ḫt* 7 instead of ending on that day. In that case it would have ended on IIII *3ḫt* 13."[60]

Both Parker and Gardiner were fixated on the key feast of *Nḥb-k3w*.[61] Both scholars tended to confuse the two overlapping religious celebrations which occurred in antiquity as has occurred in our modern era. Nevertheless, Parker was on the right track when he reiterated that *Nḥb-k3w* was set, in a civilly ordered fashion, on the first day of Mechir.[62] Examples in which *Nḥb-k3w* took place on I *prt* 1 are well-known, among them the Medinet Habu Calendar (list 52), an important reference narrated in the famous chronicle of prince Osorkon, and a few others.[63] Undoubtedly even

---

pp. 28–31. Peter Beylage follows Helck's interpretation of the sale of the office in *Aufbau der königlichen Stelentexte vom Beginn der 18. Dynastie bis zur Amarnazeit* (Wiesbaden, 2002), pp. 403–11 and 743–45 (with notes on the oracle during the Opet feast). Add now D. Lefèvre, "Egyptian Queens: A Selection of Sources," in Christiane Ziegler, ed., *Queens of Egypt: From Hetepheres to Cleopatra* (Monaco and Paris, 2008), pp. 407–8. The best and most detailed recent study is by Marcella Trapani, "Une nouvelle enquête sur la stèle d'Ahmès-Néfertari," *ZÄS* 129 (2002): 152–65.

58 "The Problem of the Month-Names: A Reply," 105.

59 O. CGC 1059 (recto, lines 7–8): Georges Posener, *Catalogue des ostraca hiératiques de Deir el Médineh* I (Cairo, 1938), pls. 32 and 32a. This is also referenced in note 88, below.

60 "The Problem of the Month-Names: A Reply," p. 105, note 1.

61 Gardiner's contribution was in his "The Problem of the Month-Names," pp. 14–16 (with his reference to the Ahmose-Nofretari Donation Stela) and 30–31.

62 I provided a detailed analysis of this material in footnote 2 to *The Private Feast Lists of Ancient Egypt*, pp. 32–33.

63 Cf. *KRI* V 172–73 for Lists 52–55; in general, see Siegfried Schott, *Altägyptische Festdaten*. Akademie der Wissenschaften und der Literatur, Abhandlungen der Geistes- und Sozialwissenschaftlichen Klasse 1950.10 (Wiesbaden, 1950), pp. 93–94 — for convenience, reference 83 is *Urk.* IV 177. Cf. Krauss, *Sothis- und Monddaten*, p. 79. One can add here the evidence from the Chronicle of Prince Osorkon, wherein the key date of V, 1 is linked with coronation. Most recently, see Roberto B. Gozzoli, *The Writing of History in Ancient Egypt during the First Millennium BC (ca 1070–180 BC): Trends and Perspectives* (London, 2009), pp. 41–49. For the evidence of *Urk.* IV 177, see Helck, "Feiertage und Arbeitstage in der Ramessidenzeit," *JESHO* 7 (1964): 139–40,

more troublesome was the connection of the Sokar Feast with these two religious occurrences, *Nḥb-kꜣw* and *kꜣ-ḥr-kꜣ*.[64]

Fortunately, it is no longer difficult to distinguish each of the three festivals. If *Nḥb-kꜣw* was, historically speaking, securely set on the first day of the second season, the famous "second *wp rnpt*," then Sokar, preceding it, originally ran from day 25 to 26 in IV *ꜣḫt*. Eventually, it expanded to include the last five days of the fourth civil month and thereby exactly paralleled the temporal interval of the epagomenals. In essence, the five days of Sokar performed the same function as those at the end of the civil year: they concluded a major defined temporal interval and, more importantly, served as the end of a cycle. Osiris, after all, is reborn on the morning of I *prt* 1.[65] Thus there was an association of all three celebrations in the minds, if not the hearts, of the ancient Egyptians.

---

and *Materialien zur Wirtschaftsgeschichte des Neuen Reiches* III (Wiesbaden, 1963), p. 353, but with inaccuracies (e.g., his "day 9" for *psḏntjw*). Nonetheless, one can refer back to the review of Parker, *BiOr* 9 (1952): 101, as he clarifies some of the inaccuracies in Sethe's copy. See now Spalinger, "The Calendrical Importance of the Tombos Stela," *SAK* 22 (1995): 271–81 and pl. 12. I wish to cite the following reference from a Cairo statue, CGC 42210, because it might be overlooked in this context: Georges Legrain, *Statues et statuettes de rois et de particuliers* III (Cairo, 1914), p. 26; and Jansen-Winkeln, *Inschriften der Spätzeit. Teil II: Die 22.–24. Dynastie* (Wiesbaden, 2007), p. 234. The text includes these important feasts: *wp rnpt*, Opet, *nḥb-kꜣw*, the king's accession (*ḫꜥ n nswt*—not understood by Legrain or by Jansen-Winkeln), Chonsu, the Valley, and Epiphi. The order courses nicely through the civil year.

64 In general, let me cite the two basic studies of the Sokar feast: Gottfried Wohlgemuth, *Das Sokarfest* (Göttingen, PhD Dissertation, University of Göttingen; 1957), and Kenneth A. Kitchen and G. A. Gaballa, "The Festival of Sokar," *Or* 38 (1969): 1–76. Add Louise Mikhail, "The Festival of Sokar: An Episode of the Osirian Khoiak Festival," *GM* 82 (1984): 25–44. See *KRI* III 457.12 for an excellent case, albeit fragmentary, of offerings set on IV *ꜣḫt* 26, and the feast of Sokar must be intended.

65 The crucial importance of I *prt* 1 for the heb sed was discussed by S. Voß, *Untersuchungen zu den Sonnenheiligtümern der 5. Dynastie: Bedeutung und Funktion eines singulären Tempeltyps im Alten Reich* (Hamburg, University of Hamburg, Ph.D. dissertation, 2004), pp. 124–34. Add, nonetheless, my "Calendars: Real and Ideal," in Betsy M. Bryan and David Lorton, eds., *Essays in Egyptology in Honor of Hans Goedicke* (San Antonio, 1994), pp. 297–304; "Parallelism of Thought," in Catherine Berger, Gisèle Clerc, and Nicolas Grimal, eds., *Hommages à Jean Leclant* IV (Cairo, 1994), pp. 363–77; and my "Chronological Remarks," p. 56. Most certainly the "second" *wp rnpt* of V 1 needs more research. Erik Hornung and Elisabeth Staehlin, *Neue Studien zum Sedfest* (Basel, 2006), pp. 39–40, provide the necessary background data concerning the date of V 1 and the heb sed festival. Additional comments will be found in Chapter 1, note 65.

Civilly speaking, the event of Sokar was originally set on 25–26 of IV *ꜣḫt*. I will not attempt to unravel its pre-civil orientation, except to state that it probably should have always taken place before the commencement of the second season.[66] The expansion of the duration of Sokar meant, in time, that its original lunar orientation was forgotten or, at best, dimly perceived. Ulrich Luft has observed that, based on the Twelfth Dynasty material in the Illahun archive, the evening of the beginning on IV *ꜣḫt* 25 is recorded as well as the expected date of IV *ꜣḫt* 26.[67] He felt that no specific "Mondtag" could be assigned to the festival's commencement. On the other hand, since the ceremony was of extreme antiquity, it could only have been regulated by some natural occurrence. Since "predicting" the five days before the archaic second season would have been difficult, I posit some type of lunar influence. By the Middle Kingdom, notwithstanding the Illahun evidence, one clear reference is to IV *ꜣḫt* 14, an anomaly that Luft recognized.

In historical times, changes are observable. The evidence of Gaballa A. Gaballa and Kenneth A. Kitchen from the Medinet Habu Calendar establishes that by the reign of Ramesses II, Sokar conveniently ran for one Egyptian week (10 days).[68] It began on day 21 of the fourth civil month and concluded on the last day of that month.[69] The two scholars reflected upon the familiar connections to rebirth and rejuvenation, as well as to the sun god Re. At IV *ꜣḫt* 30, just before the next season, the "Raising of the Djed Pillar" marked the impending rebirth of Osiris. In later times, the religious festivities grew even longer in duration, as both P. Louvre N 3176 (S) — from day 18 to 30 in the fourth civil month — and the Dendera evidence — encompassing IV *ꜣḫt* 12 to 30 — indicate. Indeed, the name of the event was eventually changed to Choiak, after the feast of *kꜣ-ḥr-kꜣ* and Sokar, as a designation, eventually succumbed to desuetude.

An original northern, Memphite-based, mortuary cult had expanded over millennia to include southern, Theban, aspects as well as the Osirian rites of Abydos. Kitchen and Gaballa remark that the ceremonies related to the erection of the djed pillar were unknown in the Middle Kingdom, whereas they were prominent by the Eighteenth Dynasty.[70] Hence, at the latest by that time, the rise of Thebes affected the prolongation of the Sokar festival as well as its very nature. But the important thing to remember is that this religious event lost its circumscribed archaic orientation during the millennia of pharaonic Egypt.

66 See Ulrich Luft, *Die chronologische Fixierung des ägyptischen Mittleren Reiches nach dem Tempelarchiv von Illahun*, pp. 179–81; and Spalinger, *The Private Feast Lists of Ancient Egypt*, chapters 1 and 2.

67 Ibid.

68 "The Festival of Sokar," pp. 34–48 and 71–74.

69 *KRI* V 161.3–72.8 (Lists 42–51). Note that the last day was *not* plastered over by the new festival of the Meshwesh victory. It was never touched.

70 Ibid., p. 72.

An excellent parallel can be seen in the Theban Opet festival. There, the Eighteenth Dynasty date of II *ꜣḫt* 15 undoubtedly indicates an original lunar-based ceremony set upon the half moon of the lunar month.[71] The Medinet Habu calendar, copied from a Ramesses II exemplar in the Ramesseum, sets Opet from II *ꜣḫt* 19 and runs it to the end at III *ꜣḫt* 12.[72] Krauss refers to one interesting piece of evidence for the event dated to Amunhotep III.[73] In this case, the association with the moon is straightforward. The later commencement on day 19, probably established so that the event would run to the end of civil month II, reveals the lunar solar epoch (30-11 = 19), just as the earlier beginning, day 15, needs no commentary. The overriding importance of this Theban celebration may be seen in the great Papyrus Harris with a timeline of II *ꜣḫt* 19 to III *ꜣḫt* 15, twenty-seven days in all, and the evidence from the two-dimensional water clock contained in Part I of the Tanis Geographical Papyrus. In the latter, the second civil month is explicitly called [*P(ꜣ)* ?-]*Jpt*, what the New Kingdom Egyptians probably very early called Phaophi.[74]

Perhaps my divergence from the main theme of this presentation seems to be more extensive than some had expected. Nevertheless, this survey has had as its purpose the goal of addressing other ways that the Egyptians could adapt their newly-created Civil Calendar to the lunar-oriented feasts that were the basis for all previous cultic (or other) religious occurrences which were in some way repetitive. (In the aforementioned example, annually.) We can now see that the eponymous feasts operated differently from Opet or Sokar. The same may be said with regard to the celebration of Thoth, which was firmly established to be on day 19 of the first civil month.[75] Therefore, let us return to the second of the three overlapping events.

That the permanent setting of *Nḥb-kꜣw* was on the first day of the second season, I *prt* 1, leads one to suspect that it probably was oriented to the first day of the second lunar season or, equally, to the first day of the fifth lunar month. Because I have covered the controversy over *Nḥb-kꜣw* between Parker and Gardiner, I will

---

71 For the chronological backdrop: *Urk*. IV 722.1 (with the broken reference to ⸢II⸣ *ꜣḫt* 13 + x and 824.10 (II *ꜣḫt* 15, and lasting eleven days). Cf. Gardiner, "Tuthmosis III Returns Thanks to Amun," *JEA* 38 (1952): 22, with note 4; Schott, *Altägyptische Festdaten*, pp. 84–85; and Spalinger, "The Calendrical Importance of the Tombos Stela," p. 294, note 16.

72 *KRI* V 148.13–56.12 (Lists 29–35).

73 *Sothis- und Monddaten*, p. 152 and note 2 = William C. Hayes, "Inscriptions from the Palace of Amenhotep III," *JNES* 10 (1951): 138, fig. 11—*rnpt sp* 35 II *ꜣḫt sw* 19 *ꜥḏ wꜣḏ n ḥb Jp*⸢*t*⸣.

74 P. Harris' evidence (pl. 17a 5) is discussed by Pierre Grandet, *Le Papyrus Harris I (BM 9999)* (Cairo, 1994), I p. 246, II p. 89. The second datum is in Francis Ll. Griffith, *Two Hieroglyphic Papyri from Tanis* (London, 1889), pl. IX (Part I, 1).

75 Spalinger, "Thoth and the Calendars," in Spalinger, ed., *Revolutions in Time*, chapter 3.

not retrace the secondary literature.[76] It was an extremely archaic festival that was performed in honor of the second season. Its agricultural origins remain vague, at best. It turns up in the official calendars that the pharaohs established in order to regulate their temple income, offerings, and the like.[77]

But Gardiner, as Parker, conflated this event with *k3-ḥr-k3*, even though the former raised significant and, in truth, effective refutations against Parker's conclusions. He also placed emphasis upon *Nḥb-k3w* as a day "upon which the new monarch was invested with his royal attributes."[78] Gardiner also had the advantage in the dispute of soliciting two expert Egyptologists to his side. Jaroslav Černý, the first, referred him to "a half-forgotten notebook of transcripts of Berlin ostraca," in which one, P. Berlin 12635, provided information on the subject. He translated two key lines, 9–10, as follows: "Again, first month of Winter, day 2, in the Khoiak feast, he ate one large cake (*sᶜb*) and 2 (pieces of) dressed (*dr*) meat."[79] (The original is in italics.) In other words, on day 1 of the fifth civil month, the festival of *k3-ḥr-k3* was recorded.

Yet, I do not think that there is anything abnormal or outstanding about the reference. It fits perfectly within our contentions specified in detail above. If this festival were based on the moon, then its first day would be sighted by the absence of a crescent. This was on *psḏntjw* and, furthermore, *psḏntjw* of month *K3-ḥr-k3* was its original commencement. When the Civil Calendar swept the less efficient system aside, the event easily was placed within that same-named month, but owing to the *décalage* of the month names — minus one — it now started in civil month IV. The observance could start on any of the thirty days of this civil month IV, and it might have lasted up to seven full days. Hence, Parker's evidence from the Ahmose-Nofretari Donation Stela, in combination with Ostracon DEM 1059, can be accommodated within my analysis with no difficulty. Even more, the third piece of evidence, presented by Gardiner but discovered by Roberto Caminos, provides additional support for this new interpretation. The Tanis Geographical Papyrus indicates that on day 1 of civil month V, the feast of *k3-ḥr-k3* was celebrated.[80] To Gardiner, this was a very

---

76 "The Date of Amunhotep II's First Accession," *SAK* 40 (2011): 387–97.

77 Two key examples are *Urk.* IV 177 (from Karnak) and *KRI* V 172.9–173.12 (List 52 of the Medinet Habu calendar). The latter might do "double service" for the sake of this argument because Ramesses III utilized the archival copies of the Ramesseum for his festival calendar.

78 "The Problem of the Month-Names," p. 28.

79 Ibid., pp. 30–31; the amusingly-written quote will be found on page 30. The ostracon under consideration is O. Berlin 12635, and I must thank Dr. Jaromir Malek for allowing me to see the Černý transcription. It may be found in Notebook 32, pp. 44–45. The key reference is on page 45.

80 Griffith, *Two Hieroglyphic Papyri from Tanis*, pl. XII (Part 7, 37–38 for full details). I have collated this portion of the papyrus with the courtesy of Dr. Richard Parkinson; see Gardiner, "The Problem of the Month-Names," p. 31.

powerful piece of evidence that refuted Parker's hypothesis. The papyrus also states: "IV *ꜣẖt* [30 (?)]: Erection of the Djed Pillar (?)," followed by "I *prt* ⌈1⌉: Feast of *kꜣ-ḥr-kꜣ*."

Again, I see no contradiction to my reconstruction of the event. It is lunar, as Parker realized, but in a way different from his conclusions. Gardiner, on the other hand, would perhaps feel that the final *pièce de resistance* that was added to Chapter 42 of the Book of the Dead could be used to support his dating.[81] It actually cannot. The passage in question refers to the fifth civil month and day 4. Gardiner read: *tpj prt sw* 4 *n jrt kꜣj-ḥr-kꜣ m tpj <n> Nḥb-*⌈*kꜣw*⌉, whereas Parker preferred: "being I *prt* 1, the fourth day of celebrating the Khoiak-feast and the first of *Nḥb kꜣw*."[82] As we have no way to refute the equation of I *prt* 1 = day of *Nḥb kꜣw*…, of the two interpretations this contention has the greater strength. But even if Parker is followed, the determination of the sighting of the moon will still remain troublesome for Parker.

Let us follow the newer "rule" that I have argued best explains the transference of months names from the original lunar calendar to the later civil one. Irrespective of *Nḥb-kꜣw*, *kꜣ-ḥr-kꜣ* must have been the name of the celebration set on the first day of the original fifth lunar month of the old Egyptian lunar year. The name moved back into the fourth position with the inception of the Civil Calendar, but the festival kept its dating in that *kꜣ-ḥr-kꜣ* still began on lunar day 1. Because it had to take place in the month of *Kꜣ-ḥr-kꜣ*, the day of *psḏntjw* would have been in civil month IV because the name of the latter was now *Kꜣ-ḥr-kꜣ*. If Parker's interpretation of this Book of the Dead papyrus is preferred, then on one occasion the beginning of the feast of *kꜣ-ḥr-kꜣ* occurred in civil IV *ꜣẖt* and on day 28. Similarly, the Ahmose-Nofretari Donation Stela located the same celebration on day 7 of the fourth civil month. Once more, we can present the case that *kꜣ-ḥr-kꜣ* commenced either on that day or somewhat earlier.[83] The new piece of information brought into the calendrical court by Parker — namely, Ostracon DEM 1059 — fits perfectly within this new analysis. Regardless of the length of this religious event, the sighting of *psḏntjw* could only have occurred within civil month IV. There is no need to hypothesize a later or "second lunar calendar," as Parker did. This feast, as all others based on the moon, thus "bobbed" back and forth in a situation akin to that proposed by Parker in his *Calendars*.[84]

As a final example of lunar festivities and their chronological placement within a civil setting, Epiphi is also a very useful, indeed paradigmatic, case because it is the name of the eleventh civil month.[85] Gardiner found an example of an oracle inscrip-

81 "The Problem of the Month-Names," p. 15.

82 "The Problem of the Month-Names: A Reply," p. 105. His major point was that the sun disk occurs twice. He italicized the last word after *prt* and the first word before the four strokes for "four."

83 Ibid., p. 105, note 1.

84 Pp. 53–54.

85 Parker, *Calendars*, pp. 46 and 58–59; and Krauss, *Sothis- und Monddaten*, pp. 151–54.

tion which supposedly refuted Parker.[86] Naturally, there was a riposte by Parker, but he was forced to admit that these eponymous festivals took place "on the first day of the month." (I do not insist that the commencement of any of these religious events had to be dated to day 1 of a *civil* month.) Krauss added more data to that brought forward by Parker and, although one slip of his can be noted, his summary is worthwhile to review.[87] Here are his four crucial dated examples of the commencement of the Epiphi feast. Note that we do not know how long it lasted.

Civil Month XI 28
Civil Month XI 29
Civil Month XII 1–2
Civil Month XII 2

At the minimum, we can see that it could begin close to the end of the month of Epiphi. Moreover, it lasted for two days at the minimum. Thus, it could fall within the final civil month and on the first two days. In essence, I see nothing contradictory to the principles presented above. The Epiphi feast, which gave its name to the original twelfth lunar month or, to be more precise, the first day of that event, had to fall in civil month XI. There is no need for concern that the Epiphi feast, or any other eponymous event, had to begin on a day 1 of a civil month. It is unfortunate that other data from Deir el Medineh do not help our cause.[88]

86 Gardiner, "The Problem of the Month-Names," p. 12, following Nims, "An Oracle Dated in 'The Repeating of Births'"; now see *KRI* VI 702.5–703.3.

87 *Sothis- und Monddaten*, p. 151. On the same page he refers to Černý, Notebook 17, p. 175. This is P. CGC 58096, which I have consulted thanks to the courtesy of Dr. Jaromir Malek; see now *KRI* VII 102–103.10. The date is regnal year forty-three of Ramesses II. That jewelers' account contains a date of day 16, and not 15 as Krauss reads, in the month of Epiphi. Note that a civil month can be referred to by name, a reasonably well-known situation. For easy accessibility, the other Ramesside cases may be found in: *KRI* IV 218.14–15 (Krauss' note 7 = O. DEM 209), and *KRI* VI 697.8 (Krauss' note 8 = Botti-Peet, *Giornale*, Tav. 59). For the remaining case on the Banishment Stela, Louvre C 256, see Kruchten, *Le grand texte oraculaire de Déhoutymose*, pp. 24–27 and 310–21; with the older study of von Beckerath, "Die 'Stele der Verbannten' im Museum des Louvre"; and now Jansen-Winkeln, *Inschriften der Spätzeit. Teil I*, pp. 72–74.

88 R. Demarée, "Month Names and Feasts at Deir el Medîna," in R. J. Demarée and Jac. J. Janssen, eds., *Gleaning from Deir el-Medîna* (Leiden, 1982), pp. 193–44. I have systematically checked all of the unpublished material in this study as it relates to feasts. O. DEM 1059, referred to earlier (see note 59, above) is listed on page 221. Cf. the attempt to understand the festival situation in the workmen's village by Heidi Wikgren, "The Festival Calendar at Deir el-Medina," in Kathryn Piquette and Serena Love, eds., *Current Research in Egyptology 2003* (Oxford, 2005), pp. 179–200 and especially pp. 184–85.

The interpretation offered here has the distinct advantage of being simple and straightforward. In historical times, and after the introduction of the Civil Calendar, lunar sightings still occurred. They were, however, now embedded within the new "rational" Civil Calendar. One either waited for a specific lunar day or noted its presence. A useful and historically early example of this may be seen in an Eleventh Dynasty stela from Coptos that refers to the fifteenth lunar day occurring on day twenty-five of the fourth month of Achet.[89] But no "second lunar calendar" was ever in existence.[90]

89 The references are supplied in note 69 to Chapter 1.

90 As a fitting conclusion, I can do no better than refer the reader to the analysis of Jean Winand, "Un mystère enfin éclairci!," in *"Wo ist denn hier der Satz?" Fastschrift für Friedrich Junge überreicht zu seinem 60. Geburtstag am 18. April 2001* (= *Lingua Aegyptia* 9 ½ [2001]).

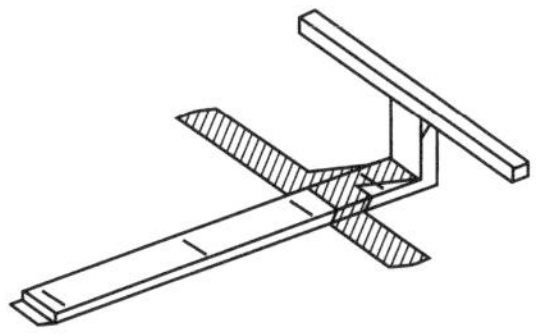

# 4

# Mathematical Factors of the Battle of Kadesh

The temporal nature of events, when presented in a simple fashion and limited to one major encounter every day, can provide the historically minded onlooker with a gamut of possibilities. It is of particular interest to us that historians of war, when limiting themselves to battles, are keen to specify the progress of a battle and the constraints of natural occurrences, especially sunrise and sunset. In addition, to be able to reconstruct the phases of one mighty armed clash provides us with an evolutionary perspective owing to the emphasis upon the diachronic nature. Time thus forces the modern viewer into a detailed, almost blow-by-blow reconstruction of the military encounter. It is for this reason that detailed reports, no matter how biased or one-sided they may be, are greatly appreciated by historians and enthusiasts alike. Not surprisingly, the two major battles of the Egyptian New Kingdom, Megiddo under Thutmose III and Kadesh under Ramesses II, have been at the focus of voluminous historiographical research.[1]

---

1 It is not surprising that the basic study with key logistical considerations was presented by James Henry Breasted, "The Battle of Kadesh: A Study in the Earliest Known Military Strategy," in *Decennial Publications of the University of Chicago* V (Chicago, 1903), pp. 81–126, and it still remains the crucial stepping stone for all research. As the bibliography is voluminous, let me single out some major detailed compendia: Kenneth A. Kitchen, *Ramesside Inscriptions: Historical and Biographical. Notes and Comments* II (Oxford and Malden, 1997), pp. 3–55; Anthony Spalinger, *War in Ancient Egypt* (Malden, Oxford, and Carlton, 2005), chapter 13 and pp. 209–26 in particular; Giacomo Cavillier, *La battaglia di Qadesh: Ramesse II alla conquista dell'Asia, fra mito, storia e strategia* (Turin, 2006); and Michael G. Hasel, "The Battle of Kadesh: Identifying New Kingdom Polities, Places, and Peoples in Canaan and Israel," in S. Bar, D. Kahn, and J. J. Shirley (eds.), *Egypt, Canaan and Israel: History, Imperialism, Ideology and Literature* (Leiden and Boston, 2011), pp. 65–85. Carola Vogel's recent "Die Schlacht von Megiddo: Thutmose III. im Kampf um Syrien und Palästina," *Ished* (Journal des Ägypten Forum Berlin e.V.) 1/2001: 15–28, can be added to the bibliography in the basic volume of Ronald B. Redford, *The Wars in Syria and Palestine of*

In essence, parameters involving mathematical evaluations normally cover the logistic nature of battles and campaigns. One expects to be able to reconstruct the time elapsed between various localities, such as staging points and cities. In addition, the rate of advance (or even retreat) is important to know, especially because we need to know why certain battlefield encounters occurred at position X instead of, for example, the desired terrain of position Y. Strengths of the opposing forces, latent or involved, are equally significant for any historical reconstruction that desires to achieve a close approximation to reality. But by no means does a purely mathematical approach solve the broader ramifications of warfare "as it was." If troop size is not ignored, is morale? If the technological ability of one side has been sufficiently explicated, does this imply that we can statistically weigh the probabilities for ultimate success of both protagonists? Clearly, much more is needed than a single-minded adherence to arithmetic, but I believe it fair to state that without a relatively complete list of measurable parameters, it is impossible to reconstruct the outcomes of military conflicts very well.

The positions of the Hittite forces vis-à-vis the Egyptians can be listed, especially in the light of most recent scholarship.[2] Note that I have followed, in general, Breasted's reconstructions and the later "corrections" of Major Alfred H. Burne, although recent studies have tended to re-evaluate these now dated studies.[3] Yet, it must be stressed that the latter correctly separated the chariot sector of the opposing Hittite army from the infantry one. No doubt, Muwatallis planned this in advance, although we do not know the time it took to achieve. I suspect that Muwatallis planned his surprise rapid movement of the war vehicles well before Ramesses moved with his advance division of Amun out from the "Forest of Lebawi."[4]

---

*Thutmosis III* (Leiden and Boston, 2003). Most recently, see Frédéric Servajean, *Quatre études sur la bataille de Qadech* (Montpellie, 2012). I see no reason to change the positions advocated in this study, one written before Servajean's analysis appeared.

2 Jürgen Kenning, *Der Feldzug nach Kadesch: Das Ägypten des Neuen Reiches auf der Suche nach seiner Strategie* (Hildesheim, 2014), passim, but especially pp. 134–38; and Claude Obsomer, "La bataille de Qadech de Ramsès II: les *n'arin, sekou tepy* et questions d'itinéraires," in Christina Karlshausen and Claude Obsomer, eds., *De la Nubie à Qadech/From Nubia to Qadesh. Études sur la guerre dans l'Égypte ancienne* (Brussels, 2016), pp. 81–170.

3 This old study is useful to cite as well, owing to its importance: Alfred H. Burne, "Some Notes on the Battle of Kadesh: Being a Military Commentary on Professor J.H. Breasted's Book, 'The Battle of Kadesh'" (University of Chicago Press, 1903), *JEA* 7 (1921): 191–95.

4 Classically, Bernard Maisler (= Mazar), "Meḥqarîm topograpiyyîm: 5. L$^{e}$bô' Ḥ$^{a}$mat = Lebo Hamath and the Northern Boundary of Canaan," *Bulletin of the Jewish Israel Palestine Exploration Society* 12 (1946): 91–100 (English summary on page VIII)—later translated, expanded, and easily available in Mazar, "Lebo-*ḥamath* and the Northern

To effect the circling motion of chariots, after they traversed the ford of the river Orontes, we have to allow for about 1 kilometer.[5] Most reconstructions of the battle do not disagree with one another at this point, if only because Breasted's visual positioning of the military forces has, in general, been followed. That is to say, there is a scholarly consensus at this point. Burne, unlike Breasted and others — particularly Kitchen — placed the second division of Pre just across the brook el-Mukadiyeh. He also situated Muwatallis' host northeast of Kadesh, whereas Burne preferred it to be adjacent to an "invisible" middle horizontal line bisecting the mound of Tell Nebi Mend itself. There, Burne felt, was the infantry with the chariotry separate and to the south of the footsoldiers. On the other hand, Claude Obsomer's recent reevaluation of the situation, especially with regard to "Kadesh the Old," needs to be taken into consideration.[6] His reconstruction would enable us to add, at the maximum, minutes onto the arrival of the Hittite chariots at Ramesses' camp.

At the minimum, we can increase the distance which the Hittite chariotry had to cover up to ca. 1.5 kilometers, owing to the arcing vector of those forces, as well as the attack at the southeast corner (and above) of Ramesses' camp.[7] The pictorial records of the accounts indicate a rather peaceful camp being invaded by the enemy. Two temporal situations have been combined in order to provide dramatic intensity to the viewer.[8] But it is the peaceful, indeed tranquil, setting that should put us on alert. Was this mere window dressing? That is to say, had the Egyptians truly settled down for some time when the Hittites attacked, or was the interval temporally short? Considering the importance of the inserted Bulletin, the equivalent of a *Königsnovelle*

---

Border of Canaan," in Shmuel Ahituv and Baruch A. Levine, eds., *The Early Biblical Period: Historical Essays* (Jerusalem, 1986), pp. 189–202; and Robert North, "Phoenicia-Canaan Frontier Lᵉbô' of Ḥama," *Mélanges de l'Université St. Joseph* 46 (1970–71): 71–103.

5 On these chariot tactics, see the summary of Fabrice de Backer, "Evolution of War Chariot Tactics in the Ancient Near East," *UF* 41 (2009): 29–45. There is a somewhat speculative study by Étienne de Vaumas, "Sur le mouvement tournant effectué par Ramsès II à la veille de la bataille de Qadesh," *Mélanges de l'Université St. Joseph* 46 (1970–71): 53–67. Compare this study with my "Notes on the Reliefs of the Battle of Kadesh," in Hans Goedicke, ed., *Perspectives on the Battle of Kadesh* (Baltimore, 1985), pp. 1–42.

6 "La bataille de Qadech de Ramsès II: les *n'arin*, *sekou tepy* et questions d'itinéraires."

7 Following Obsomer, the distance has to be increased, but not by much. More importantly, the velocity of the enemy chariots was great enough not to argue for any lengthy extension in time.

8 On this situation note Roland Tefnin, "Image, écriture, récit. À propos des représentations de la bataille de Qadesh," *GM* 47 (1981): 55–78; with Spalinger, "Notes on the Reliefs of the Battle of Kadesh." Add Gonzalo M. Sanchez, "Variations of Representation in the Direction of the Battle of Kadesh," *JSSEA* 31 (2004): 129–50.

account,[9] it seems reasonable to follow the Egyptian setting, namely, that the first division had already reached its camp, settled down, and prepared for the remaining three divisions to reach them when Ramesses was suddenly surprised by new data: the discovery that the Hittites were close by, to the northeast of the city of Kadesh. This report indicates that some time must have passed between the commencement of "settling down" and the later discovery of the "betrayal."[10] The pictorial record indicates this. Ramesses' camp was well-stocked with food and war material. All appears to have been set out in an orderly fashion. In the reliefs, fodder for the horses and oxen is present, a personal guard protecting the pharaoh is already in action, and the horses are lined up separate from their vehicles. Indeed, the minor icons of peacefulness reflected by the quiet and calm camp life among the troops have been well-represented by the artists.

Naturally, one cannot overlook the possibility that Muwatallis' attack took place soon after Ramesses pitched camp. Nevertheless, considering his fateful counterattack and the arrival of the fifth Egyptian division (the Na'arn), it appears reasonable to conclude that perhaps one hour had passed after the Egyptians established their base of operations to the west of the city of Kadesh.[11] Thus, we have to determine how long it took the Hittite chariot division to reach Ramesses' camp.[12] We also must keep

---

9 The issue of the *Königsnovelle* in the Kadesh presentations is now presented in a forthcoming study for a Festschrift, "Divisions in Monumental Texts and their Images: The Issue of Kadesh and Megiddo."

10 For the factors surrounding this complaint, see the background data in Jan Assmann, "Krieg und Frieden im alten Ägypten. Ramses II. und die Schlacht bei Kadesh," *Mannheimer Forum*, 1983/84 (1984), 175–231.

11 For the problems of time (sunrise, sunset, and the arrival of the Egyptians), I referred the reader to pp. 43–47 of Kitchen's analysis referenced in note 1, above, in my *War in Ancient Egypt*, p. 232, note 10 (top of page). The problems of the *exact* dating of the battle, dependent upon the presumably fixed first regnal year of Ramesses II — still argued to be 1279 BC, but this is no longer certain — do not affect any temporal calculations involving sunset.

12 None of these crucial data have, as of yet, been discussed. Richard H. Beal's volume, *The Organisation of the Hittite Military* (Heidelberg, 1992), ignores these logistic and mathematical parameters owing to the limitation of the extant Hittite data. On pp. 148–49 will be found his speculation that the Hittite system of three men to a chariot altered from an original two-men arrangement. He bases this argument upon one scene from Seti I (northern exterior hypostyle Hall, west side, register two). Unfortunately, Beal does not take into consideration the method of Egyptian iconography. The Hittite chariotry is discussed on pp. 141–90. Beal contends that cavalry contingents were present in the Hittite army; I suspect not. Details of the size of the Hittite army will be found on pp. 277–96, although his analysis of the sizes of the Egyptian and Hittite forces at the battle of Kadesh are valid. Most certainly, the Egyptians did

in mind that this rapid movement was constrained by the presence of the division of Pre. In other words, it was held up for some time at the southwest of Kadesh. Burne thinks that "the rear of division Reʿ would come running back south to the third division with wild stories."[13] This presumes, of course, that those running south were on chariots. But was this the case? The Egyptian army marched on foot. Hence, any escaping troops would have to have speedily escaped the initial Hittite onslaught.

In addition, the pharaoh had already sent the vizier to speed up the division of Ptah; at least one additional group went further south. The Poem also tells us that it was while Ramesses was talking with his official that the Hittites crossed the Orontes with their war vehicles. Therefore, additional time had passed from the initial point when Ramesses found out the true geographical position of Muwatallis and before the Hittites were sent westward. Figure 7 (on the next page) reflects the presently accepted historical outline of the first phase of the battle.

The latter was no doubt contingent upon the obvious sign that the Egyptians had now realized the predicament into which they were led. After all, the vizier and other Egyptian scouts could have been spotted by enemy Hittite scouts who easily could have informed their ruler of the new disposition of the Egyptians. Surely it was the capture of those men that alerted the Hittite monarch to the necessity of moving fast against the Egyptians. But the crucial point is that Muwatallis acted, and soon after, if we follow the Egyptian accounts, the point in time arrived when Ramesses called together his officials for a camp discussion.

---

not have 47,000 men, as he claims (p. 293); nor did the Egyptians muster 41,000 men. As Beal notes on p. 291, R 43 and R 44 in the Egyptian account (*KRI* II 139.14/15 and 140.2/6) give 18,000 *thr* warriors of Muwatallis and 19,000. Did these elite troops ever fight? How were these two numbers determined by Ramesses' scribes? Trevor Bryce's *Hittite Warrior* (Oxford and New York, 2007), though aimed at a general audience, provides useful commentaries with regard to the New Kingdom military forces. Note his remarks on p. 10 concerning a "chronic manpower shortage." On the same page he, too, places their number at the battle of Kadesh at 47,500 men. While disagreeing with this contention, I remain favorable to his analysis of the Hittite chariotry during that melee: "There is no indication that the Hittites benefitted in this or in any other battle from their innovation [is this the correct word?] in chariot technology" (p. 63). Kitchen. "Memoranda on Craftsmen at the Ramesseum," in Elizabeth Goring, Nicholas Reeves, and John Ruffle, eds., *Chief of Seers: Egyptian Studies in Honour of Cyril Aldred* (London and New York, 1997), p. 176, follows Elmar Edel in briefly commenting upon the Hittite baggage-wagons designated by *ṯprt* (assumed to be equivalent to Akkadian *saparru*) that can be seen in the reliefs of this battle. But there are other interpretations, Kitchen notes.

13 "Some Notes on the Battle of Kadesh," p. 194. Burne felt that one Egyptian division was composed of chariotry being equal in number as infantry (p. 192).

The period of time that the conference took is unfortunately unknown.[14] On the other hand, Ramesses' conference was apparently still going on when the Hittite chariots, 2,500 if the Egyptian record can be trusted, sped down and across the Orontes ford and then into the second division of Pre.[15] The Bulletin also reports that those enemy chariots commenced to "hem in" the Egyptians or, to be precise, the "followers of his majesty." All present reconstructions are in agreement on this point, namely, that Ramesses rapidly moved against the Hittite chariotry.

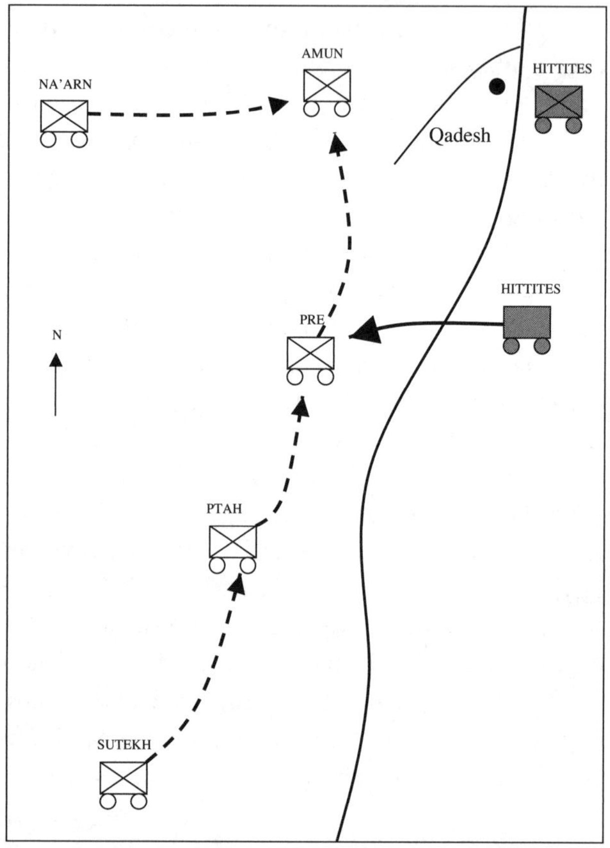


FIG. 7 The Opening Phase of the Battle of Kadesh (taken from B. Heagren, *The Art of War in Pharaonic Egypt* [PhD Thesis, Auckland University; 2010], p. 451 [fig. 6.2]).

14 Breasted states in his "The Battle of Kadesh," p. 110, that "The attack of the Asiatics, the discovery of their real position, and the dispatch of the messenger southwards, were all three practically simultaneous." Were they? And what does "practically" imply?

15 Concerning these numbers (and other closely related parameters), see my *War in Ancient Egypt*, pp. 212–17. Raymond O. Faulkner, "The Battle of Kadesh," *MDAIK* 16 (1958): 96–98 also provides useful integers of distance relating to the gap between divisions. He also thinks that "the first news Ramesses had of the disaster [= incapacitation of the division of Pre; my comment] was fugitives from the routed troops bursting into his camp with Hittite chariotry in hot pursuit." This is imaginative. The key section in the Poem is P 84 (*KRI* II 31.6/9). Hans Delbrück's classic *Numbers in History* (London, 1913) remains the best short sober account of how to understand these integers and fractions that historical accounts of war present to us. Marco de Odorico presented a volume on his specialist research, *The Use of Numbers and Quantifications in the Assyrian Royal Inscriptions* (Helsinki, 1995), one that can be consulted with profit. See, as well, Kenning's discussion in *Der Feldzug nach Kadesch: Das Ägypten des Neuen Reiches auf der Suche nach seiner Strategie* (Hildesheim, 2014).

The velocity of chariots in the Ancient Near East during the Late Bronze Age needs further examination.[16] At present, the best study remains that of Jérôme Spruytte, whose "experimental studies" on ancient harness systems remains a classic.[17] Granted that recent Egyptological scholarship has presented a more detailed and nuanced analysis of the development of the horse and chariot within pharaonic Egypt, it remains the case that our knowledge of the effectiveness of these war vehicles is often misunderstood. For example, they were most certainly not akin to modern tanks, including primitive World War I ones. In thick and high grass, chariots could not move effectively. They also needed level grounds or roadways for effective velocity. They were relatively light, and one Egyptian man could easily carry the cab plus wheels with little difficulty.[18] Finally, they often were dismantled for long foot advances. The situation of Thutmose III within the Aruna Pass is yet another case wherein the horses went single file, as did the warriors, and the chariots would have had to be taken apart and carried by men or horses at a regular pace. He encountered no obstacles, and the time he took from Gaza onwards fulfills the

16 Background data concerning these late Bronze Age chariots (Egyptian and Hittite, though we know little about the latter) may be read in: Lawrence E. Stager, "Chariot Fittings from Philistine Ashkelon," in Seymour Gittin, J. Edward Wright, and J. P. Dessel, eds., *Confronting the Past: Archaeological and Historical Essays on Ancient Israel in Honor of William G. Dever* (Winona Lake, 2006), pp. 170–76; Spalinger, "Some Notes on the Chariot Arm of Egypt in the Early New Kingdom," in Panagiotis Kousoulis and Konstantinos Magliveras, eds., *Moving Across Borders: Foreign Relations, Religion, and Cultural Interactions in the Ancient Mediterranean* (Leuven, Paris, Dudley, 2007), pp. 119–37; and Juan-Pablo Vita, "The Power of a Pair of War Chariots in the Late Bronze Age: On Letters RS 20.33 (Ugarit), BE 17.33a (Nippur), and EA 197 (Damascus Region)," in Jordi Vidal, ed., *Studies on War in the Ancient Near East: Collected Essays on Military History* (Münster, 2010), pp. 87–93. Beal, *The Organisation of the Hittite Military*, pp. 127–29 discusses a fragmentary Hittite Old Kingdom text (KBo 3.34 ii 21–35; see additional comments in his appendix 3B) in which the parameters of chariot warriors are given. Not surprisingly, this includes competence in archery as well as in horsemanship. Bryce summarizes the data in his *Hittite Warrior*, p. 27.

17 Jérôme Spruytte, "L'attelage égyptien sous la XVIIIe dynastie: étude technique et technologique," *Kyphi* 2 (1999): 77–87; *Early Harness Systems: Experimental Studies. A Contribution to the History of the Horse* (London, 1983). However, he employed one man per chariot.

18 See the key comments of Brett Heagren below in notes 19 and 22.

19 Brett H. Heagren, *The Art of War in Ancient Egypt: An Analysis of the Tactical, Logistic, and Operational Capabilities of the Egyptian Army (Dynasties XVII–XX)* (Auckland, PhD Dissertation, University of Auckland; 2010), pp. 201 and 206 in particular, and with his comments on pp. 195 and 251; Kenning, *Der Feldzug nach Kadesch: Das Ägypten des Neuen Reiches auf der Suche nach seiner Strategie*, pp. 134–38, parallels Heagren's earlier analysis.

expected distance covered of 20 km/day. Brett Heagren, on the other hand, prefers 22 km/day.[19] After taking into consideration the use of oxen with cart and donkeys as load bearers, we ought to decrease the pace.[20]

Spruytte's analyses indicated that an Egyptian chariot, carrying one man and weighing ca. 70 kg, traversing very indifferent ground for one km was able to reach a top speed of 38 km/hour (23.6 miles/hour).[21] We must, however, take into consideration the Hittite chariots. Here, despite Richard Beal's painstaking research on the Hittite army, the data remain slim. No example of a Hittite chariot has been found in the archaeological evidence. From the fact that these people used three men to a vehicle chariot, albeit with two horses, we can hypothesize a slower velocity. Moreover, Spruytte's data were for an Egyptian chariot bearing only one man. We can, therefore, hypothesize ca. 11 to 25 km/hour maximum for 2–3 men, especially if we realize that the Hittite chariots would have been stronger (heavier in mass) than the Egyptians, and so travel slower as they contained one extra man.[22]

There are three possible combinations of masses to be considered. We can eliminate the first. We shall assume that an adult charioteer weighed ca. 70 kg and a horse about 400 kg, but let me assume that the chariot weighed the same, which was not true.

a. One man/chariot: kilograms = 70 + 20 + 400 = 490
b. Two men/chariot: kilograms = 140 + 20 + 400 = 560
c. Three men/chariot: kilograms = 210 + 20 + 400 = 630

As F = ma, a = F/m. The acceleration is thus F/560 for the Egyptian side and F/630 for the Hittite. The difficulty of the joins on both sides of the chariot, where the axle was not that securely held in space, was not well-resolved at this time. We have already provided a maximum velocity for chariots with one man, namely, 38 km/hour. Assuming no impact of friction, either mechanical or by air, the maximum velocity

20 The following study may prove useful to ancient historians of war: Department of the Army, Headquarters, *Foot Marches* (Field Manuel No. 21–18), (Washington, DC, 1990). Itzick Shai et al., "The Importance of the Donkey as a Pack Animal in the Early Bronze Age Southern Levant: A View from Tell es Safi / Gath," *ZDPV* 132 (2016): 1–25, is the best study on donkeys that I have read.

21 *Early Harness Systems*, p. 49.

22 Heagren, *The Art of War in Ancient Egypt*, pp. 74–87 on chariot warfare, the crux of the matter here at Kadesh. Marian H. Feldman and Caroline Sauvage, "Objects of Prestige? Chariots in the Late Bronze Age Eastern Mediterranean and Near East," *ÄuL* 20 (2010): 67–181 discuss the Late Bronze Age chariots from a yet another useful, though not military, viewpoint. For the sake of Egyptology, I can add the following little-read study: Charles Hinson Carver, *The Egyptian Chariot* (Providence, Brown University, Ph.D. Thesis, 1955). For me, it was pleasant to see the signatures of Ricardo A. Caminos and Richard A. Parker.

should remain constant. The only change in performance of the chariot would be an increase in time to attain same maximum velocity as with one man instead of two or three. To put it another way, and one more dynamic, because F = ma, a will decrease as we increase the mass. As m increases, F should, however, not change. The maximum velocity simply will take longer to attain.[23]

For these reasons the time it would take Muwatallis' chariots to reach their maximum velocity at, say, ca. 38 km/hour, would be longer than if only two men were positioned on the chariot. This would only matter when the force reached the second division of Pre. At that point, given the imponderabilites of the encounter, we have to place the distance from the camp around 2 km. The verb employed in the account is "to cut through" or "to slice through," *šʿf*, whereas the Bulletin employs the verb *ʿq*, "to enter."[24] The Poem also notes that these soldiers were marching and were not "prepared for fighting." The Bulletin reiterates this analysis.

Thus, if there had been no obstacle preventing the Hittite chariot division from reaching the Egyptian monarch at his base camp, we can say that it would have taken about 6 minutes for them to have reached Ramesses. Allowing for indeterminate

23 Spruytte's later study, *Attelages antiques libyens: archéologie saharienne expérimentale* (Paris, 1996) is a more up-to-date study of the velocity, and its effects, of these military vehicles. His comments with regard to an Asia Minor origin for the chariotry system of Libya during the first millennium BC may be questioned. After all, despite Sea People contacts with the coastal regions of Libya during the Ramesside Period, the Egyptians themselves must have contributed much to this new military technology in the west. This is especially important to emphasize, if only because of the close Libyan (Meshwesh/Libu) interaction with the Delta and the Nile Valley from Dynasty XX onwards.

24 Poem—*KRI* II 26.12/16 (*šʿf*; P 72); Bulletin *KRI* II 118.7/10 (*ʿq*; B 79). Gardiner, *The Kadesh Inscriptions of Ramesses II* (Oxford, 1960), p. 8 has "broke into (?)" for P 72; he briefly discusses this on p. 18, noting its presence in R 11 (*KRI* II 132.6/10, with lacunae) as well. Breasted, "The Battle of Kadesh," p. 109, has "cut through," and this I accept. Among others, see Thomas von der Way, *Die Textüberlieferung Ramses' II. zur Qadeš-Schlacht* (Hildesheim, 1984), p. 299, prefers "Sie griffen die Division des Pre an," but this is very weak; Faulkner, "The Battle of Kadesh," p. 103, has "they charged"; Miriam Lichtheim, *Ancient Egyptian Literature* II (Berkeley, Los Angeles, and London, 1976), p. 64, translates "attacked" (again weak); and Gerhard Fecht, "Das 'Poem' über die Qadeš-Schlacht," *SAK* 11 (1984): 290, "durchstießen" (excellent). Are these not interesting attempts, close or not, to convey the severing or splitting (and not destroying, killing, etc.) of Ramesses' second division? Cf. Dimitri Meeks, *Année Lexicographique: Égypte ancienne* III (Paris, 1982), pp. 285 and 289; Rainer Hannig, *Großes Handwörterbuch Ägyptisch–Deutsch (2800–950 v. Chr.): Die Sprache der Pharaonen* (Mainz, 1995), pp. 807 ("einbrechen," copying Gardiner et al.) and 817. Note that I do not assume that the Hittite chariots spent much time mopping up the soldiers of the second Egyptian division. This can be queried, though I have not see any major demonstration of its factor in the time of the arrival of Muwatallis' chariots to the Egyptian camp.

factors, I would heighten this time to 10 minutes total. Yet the Hittite deployment depended upon swift movement and concomitant rapid success. It was fortunate for Ramesses that he had enough time to send his vizier and other individuals south and to prepare for battle before his camp was attacked. We must remember that it is not so easy to slice through a marching (or even fixed) army division. The success of the enemies still took time. Hence, I have presented these calculations in order to establish some logistic impediments to the chariot advance of the Hittites. The core issue here is just how resilient and cohesive was the second Egyptian division. Any groups of soldiers can "slice through" an opposing side, even if both are composed of infantry alone. Unfortunately, the morale and effective arrangements of the division of Pre are not known to us.

Nevertheless, the Hittite forces reached his carefully prepared bivouac without opposition.[25] The Bulletin specifically indicates that Ramesses was still in conversation with his high officials. The time between the capture of the Hittite scouts and the attack of the enemy chariotry at his camp was therefore short. The relief caption R 11, which is the famous one that refers to the arrival of the Na'arn, also mentions that Ramesses "sat alone" when the enemy host of chariots was pressing into his camp.[26] It also states that the first division of Amun had not yet finished pitching the camp. In other words, Muwatallis' thrust westward occurred virtually soon after his scouts were caught by the Egyptians. More importantly, because the Egyptian war record indicates that Ramesses' troops had not yet completed settling down in his camp — even if the pictorial record indicates a different picture — I contend that they would have been better prepared to meet the Hittite onslaught than if they had been totally at rest and free from care. I.e., they were not "caught with their pants down."

Breasted maintained that after the Hittite chariots had effectively penetrated the second division, "the infantry and chariotry of his majesty fled before them."[27] Apparently, the division of Pre could not show much resistance. Breasted further observed that at this point in time an unknown "messenger sent by some officer of the division of Re must have reached Ramses after his dispatch of the messengers to the two southern divisions, else he would not have been obliged to torture the Asiatic scouts in order to learn the location of the enemy."[28] Most of this conclusion is reasonable and, in fact, fits perfectly into the two part division of the pharaoh's predicament; namely, his learning of the Hittite camp, much to his surprise, so to

---

25 Let us not forget the presence of Ramesses' pet lion in his camp; cf. E. van Essche, "Les félins à la guerre," in L. Delvaux and E. Warmenbol, eds., *Les divins chats d'Égypte. Un air subtil. Un dangereux parfum* (Leuven, 1991), pp. 31–48.

26 *KRI* II 131–33.2.

27 Moreover, according to Breasted, "The Battle of Kadesh," p. 111, the second division was within one half hour of Ramesses' camp. This assumes, of course, a marching distance.

28 Ibid.

speak. Subsequently, Ramesses then received news of the collapse of his second division. I part with this interpretation, as most do, by referring to the scene of bastinadoing. That depiction is included within the large snapshot conference image in the pictorial accounts. Nonetheless, the second phase of the king's camp preparation included the subsequent report of a soldier or soldiers which indicated the attack on the Pre cohort. We must also remember that some of that division — footsoldiers and chariotry — managed to reach the camp of their monarch before the Hittites did.

We can make some more analyses of the velocity of soldiers that are appropriate to this stage of the battle. The troops escaping the Hittite onslaught must have been in the vanguard of the second division. I will place the running pace on the Second Empire's standard: 1 meter and not on the Third Republic's 80 cm.[29] Here I can also refer to the Old Prussian, pre 1871, parameters of 165 to 175 paces per minute. The maximum distance clocked with men, at a run with men carrying equipment, turns out to be 350 paces per minute. And with a march column, the maximum distance covered would be, with equipment, 2 minutes or 300 to 350 paces. All of this data, I hasten to add, is based on the Prussian double-time of 165–75 paces per minute and therefore 6 minutes to the kilometer. Using this information, we can assert that if the vanguard of Pre was around two (or slightly less) kilometers from Ramesses' camp, then the time it would have taken those fleeing footsoldiers to arrive at their king's location would have been about 6–12 minutes maximum, with the chariots reaching camp earlier, of course.

This tentative conclusion allows us to maintain that the Hittite chariotry, which arrived later, spent more time in covering the distance from the division of Pre to the Egyptian bivouac. As indicated earlier, if there had been no opposition to them, then these enemy vehicles would have operated at ca. 25 km/hour in velocity and so for ca. 5 km maximum 60/5 or twelve minutes. The fleeing Egyptian chariots of Pre would have taken 60/25 or about 2.4 minutes. Whatever delay caused the advance of the enemy chariot force against the Egyptian second division, it is clear that the Egyptian account rings true 100 percent. I believe that Ramesses had enough time to prepare himself for battle, but was unable to marshal *all* of his troops for an immediate counter-attack. Thus arises, for us as well as for Ramesses, the key question: how much time did he and his troops have to face, effectively and concertedly, the Hittite chariot attack? We do not know. But his claim to have "fought alone" might indicate that, at least at first, he was bereft of a large number of his troops in his camp.

Compounding the difficulties of the Egyptians was the fact that the Amun division was not fully prepared for battle with the Hittite chariots. R 11 indirectly supports this assertion, because it places emphasis upon the effectiveness of the Na'arn division. The latter was able to "break into the host" of Muwatallis and not the king's

29 These data I had not assembled earlier: Hans Delbrück, *Warfare in Antiquity*, Walter J. Renfoe, Jr., trs. (Lincoln and London, 1990), pp. 84–85, and see p. 90, note 1.

first division. Hence, if we consider the Egyptian accounts to be valid and basically reflecting the temporal validity of the battle up to this point, then Ramesses was truly bereft of much support from his own troops.

But how long was the Egyptian second division and, in particular, how many columns made up that body of men? The Abydos reliefs are most pertinent at this point. The footsoldiers are depicted marching and protected on all four sides by groups of three footsoldiers who are just inside of the chariots and horses.[30] The additional marching warriors seem to be composed into small units of three or four men. It seems to be the case that as soon as the division moved into the fray, these men were quickly dispersed by the charioteers who rapidly sped into the Egyptians. At the rear of these men will be found at least four rows of protective chariots and men; the same may be said for the front. In addition, there are fifteen marching footsoldiers per column.

Naturally, we cannot depend upon verisimilitude in any of these representations. The Luxor pictorial evidence of L3, for example, indicates seven separate protective chariots at the rear of the footsoldiers.[31] Inside, fifteen men can be seen operating in each row. But the pictorial account also presents additional warriors marching upon the terrain and in front of these infantrymen (fifteen per row), and at least twenty-seven men can be seen in the fragmentary reliefs. Let us allow for at least fifty footsoldiers per squadron, a conclusion that Alan Schulman reiterated in his study of the New Kingdom military,[32] but with one added qualification: we are dealing solely with the Ramesside military system and not with the earlier, Dynasty XVIII arrangements. At the Ramesseum (R1 version), for example, there are at least nine footsoldiers in a column and at least six columns, but ten if not more additional columns can be argued.[33] In this scene, the Egyptian division of Amun is shown arriving at the camp preceded *and* followed by charioteers. Clearly, the group of men marching as a unit

30 Conveniently, Susanne Constanze Heinz, *Die Feldzugsdarstellungen des Neuen Reiches. Eine Bildanalyse* (Vienna, 2001), p. 282 (II.1). See as well Spalinger, "The Battle of Kadesh: The Chariot Frieze at Abydos," *ÄuL* 13 (2003): 163–99. It is not out of the way to refer the reader to Peter Raulwing and Juliet Clutton-Brock, "The Buhen Horse: Fifty Years after its Discovery (1958–2008)," *JEGH* 2 (2009): 1–106, where a very good bibliography is present concerning horses in the Ancient Near East and specifically in Egypt. Add the small but yet useful analysis of Joseph Daborn, *The Military Use of the Horse in the New Kingdom* (Swansea, University of Wales, Undergraduate Dissertation, 2005). Earlier analyses of a logistic and physiological nature will be found in my *War in Ancient Egypt*, pp. 8–12, and Heagren, *The Art of War in Pharaonic Egypt*, pp. 179–82 on fodder for animals.

31 Heinz, *Die Feldzugsdarstellungen des Neuen Reiches*, p. 289 (IV.4).

32 Alan Richard Schulman, *Military Rank, Title and Organization in the Egyptian New Kingdom* (Berlin, 1964), p. 26 (for "company," *sꜣ* = 50 men), and especially p. 30.

33 Heinz, *Die Feldzugsdarstellungen*, 291 (V.1).

with the accompanying chariot support forms a unit greater than a platoon of fifty men. At the minimum, we can argue for a company (*sȝ*) but not the entire division. Considering all the extant New Kingdom evidence concerned with "the army in formation," assembled now by Heagren, there is little doubt that the chariots, whilst advancing with riders, were protected by infantry men (and not archers).[34]

How different the two were remains a conundrum, although Jean Yoyotte and Jésus Lopez have stressed the far-ranging contrasts within the chariot sectors of the Egyptian army many years ago.[35] Andrea Gnirs later discussed similar analyses in her reevaluation of the officer class of the New Kingdom army.[36] It is unfortunately extremely difficult to qualify further or to make precise and detailed evaluations of the subdivisions within the pharaonic army and, above all, to determine exactly the size of an Egyptian "company" and the ratio between chariotry and infantry—the latter including regular footsoldiers as well as the archer divisions. For the purpose of this discussion, I shall follow Schulman and set the company at 250 men and the platoon at 50; the latter figure can be better argued than the former.

To complicate matters even further, at Medinet Habu Ramesses III's army, while setting out on his first Libyan campaign, is led by eight columns of footsoldiers, five in each line.[37] Hence, forty men made up a squad. Underneath in the same scene will be found a highly instructive reproduction of the mercenaries and other ethnic groups enlisted in the Egyptian army. Among the Egyptians, Sea Peoples (Sherden and Peleshet), Libyans, and Nubians, a system of eight men per group seems consistent with this added factor of importance: each group is physically separate from the men to the front and the rear by about one stretched pace. The latter space formed by the men would have been the norm. Marching soldiers, of course, do not advance in a file so close to one another as to be tightly knit. An additional scene from the same battle shows ten separate Egyptians moving to the left in advance of their Sea Peoples compatriots. (In this case Ramesses III has already won the battle and is returning home.)

One additional caveat in the use of this visual material must be mentioned, and it is not the expected qualifications concerned with historical accuracy and realism. The arrangement of the Egyptian army as revealed in the Medinet Habu war reliefs of Ramesses III appears different from that of Ramesses II's taken from his Kadesh reliefs. In particular, the type of warfare depended more upon the king's archer divi-

34 *The Art of War in Pharaonic Egypt*, pp. 18–43.

35 Jean Yoyotte and Jésus Lopez, "L'organisation de l'armée et les titulaires de soldats au nouvel empire égyptien," *BiOr* 26 (1969): 4–6. Their remarks concerning Thutmose IV's advance upstream are rarely discussed.

36 Andrea Maria Gnirs, *Militär und Gesellschaft: Ein Beitrag zur Sozialgeschichte des Neuen Reiches* (Heidelberg, 1996).

37 Conveniently, see Heinz, *Die Feldzugsdarstellungen*, p. 306 (I.15 = *MH* I, pls. 30–31).

sion, a fact well-known from Merenptah's war against the Libyans. (Note that one can separate the Delta battle against the Sea Peoples attack "on water," owing to the peculiar or *sui generis* encounter of the latter.) The marching Egyptian army which moved westwards into Libya under Ramesses III seems to have lacked the expected chariot support on the four sides of each subdivision of footsoldiers. In one scene we can see 64 infantrymen advancing to the left. This representation, and another relating to the land attack against the Sea Peoples, always position the marching infantry, either advancing to war or returning from the melee, without chariot support. Indeed, at Medinet Habu one depiction of the second Libyan war emphatically stressed the importance of Ramesses III's archers while at the same time noting the advance of the infantry, still without any chariot support. In a Luxor relief of the battle of Kadesh, for example, we can see three rows of charioteers advancing to the left and protecting the footsoldiers. Yet the chariots are not depicted following the soldiers, all of whom in front carry sickle-shaped swords. At best, it can be argued that the chariot sector is shown on one side (that of the viewer) and in the rear of the infantry.

Naturally, such depictions vary. At Medinet Habu, for example, various reliefs depict Egyptian footsoldiers going to war.[38] The length of a row is hard to determine, but from these cases at Medinet Habu we may hypothesize a maximum of twenty men per row. Yet from all of these representation, as well as from the physical archaeological evidence of the chariots themselves, it seems the case that the length of a chariot in addition to the yoked horses was about as long as four marching men carrying shields. Here, the metrical data coincide rather well with the pictorial evidence. The rule which I shall therefore follow is: 4 footsoldiers = 1 chariot/driven vehicle consisting of two horses plus two men.

The parameters for the division of Pre turn out to include these details:

a. 5,000 men total.
b. At least around 15 men per squad, and thus 4 chariots of protection on each side of a squad or, if we follow Schulman, a platoon.

The rate of advance is equally important to determine as is the internal arrangement with the army divisions. I have depended upon the standard assumption at ca. 15–20 km/day covered by the Egyptian army in the New Kingdom.[39] But with regard to the

38 Three key examples may be found in Heinz, *Die Feldzugsdarstellungen*, p. 306 (I.15); compare her plates on pp. 301 (I.7) and 304 (I. 12). Others could be added; e.g., the frieze of marching troops in the lion hunt scene. Note, however, that many of these depictions artistically serve as a "base sector space filler," and thus can be queried with a healthy dose of skepticism.

39 *War in Pharaonic Egypt*, p. 32; cf. Heagren, *The Art of War in Pharaonic Egypt*, pp. 338–51 and 197–98 (rate of advance with donkeys). Kenning, *Der Feldzug nach Kadesch: Das*

ranks or rows of an Egyptian army — outside of the concept of a platoon, or even a squad — it remains an open question how, *in marching order*, the footsoldiers were set up. I.e., how many men were there per rank or row? In this context, research by William Hamblin concerning the Fatimid army of Egypt may be useful to discuss.[40] He pointed out that one can establish some type of arrangement, if otherwise interpreted by others as a theoretical set-up. The argument proceeded as follows:[41]

a. If footsoldiers are marshaled in four ranks with one meter per man, one could thus mobilize 4,000 men on a front of 1,000 meters.
b. The arithmetic used by him works with one man per meter × 1,000 meters = 1,000 × 4 ranks = 4,000 meters.
c. In Hamblin's analysis, the second army, located immediately behind the first, contained three ranks. Hence, it was the sector that was the greatest. This is not evident under the ancient Egyptian military system in the New Kingdom.
d. Clearly, any attempt by an enemy observer to estimate its opponents' total size was often invalid.
e. And, following Hamblin, we can set the marching distance or space between the men at ca. 1.5 to 2 meters.

Camps are difficult to estimate in size, but for a Roman legion of 6,000 men an area of 60 acres (or .243 km$^2$) would be occupied. Additional data concerning these sizes was brought up earlier in my New Kingdom summary of the military.[42] Note that in marching, the breadth of two or more men is approximately 3 paces or .91 m and the width of a man at the shoulder is only .46 m. Naturally, the depth varies, but I had assumed .91 meters per man for a march through the Aruna Pass in the time of Thutmose III. For an army marching, we should increase this to ca. 1.5 meters. When marching, the files, however, would be closed up and I preferred .84 meter per man in any army of this time.

Hence, these are the basic parameters:

---

*Ägypten des Neuen Reiches auf der Suche nach seiner Strategie*, seems to disagree with us here.

40 William James Hamblin, *The Fatimid Army during the Early Crusades* (Ann Arbor, University of Michigan, Ph.D. Dissertation, 1985). Note his discussion on pp. 68–92 concerning numbers in the medieval war sources, especially from the point of view of the contemporaries. Sizes of armies are also considered. I have presented a different analysis of the material on pp. 214–15 of *War in Ancient Egypt*.

41 Ibid., pp. 68–71.

42 Ibid., pp. 87–88.

1. .84 m as the maximal size per man. Let me round this up to .9 m.
2. Width of a man at the shoulders is .46 m, but I now estimate 1 man = per 1 meter, as the space between men should be increased. Earlier I had placed the breadth (or width, if you prefer) of two men as .91 m.

The result is that a unit of 50 men, Schulman's "fifty-man" unit, was the basic size of any military contingent, say a "platoon." Note that he refused to contemplate the possibility of attached chariotry, basing that supposition upon an assumed clear-cut split between the footsoldiers and the elite chariotry. Indeed, his point that "the company is never associated with the chariotry" holds. We therefore can assume that a corps of warriors, consisting ideally of 500 men, was protected by chariots.

At the maximum, if 500 men marched in columns of ten men apiece and thus the width of each row would have been 50 men, the depth and breadth of such a formation would have been ca. 50 meters in breadth and for depth somewhat less than 1 m/row, say ca. .9 m. This cohort was therefore 50 m × 9 m, the total area in meters is approximately 450 $m^2$. A reasonable size for any military corps. But the size of an Egyptian cohort remains unclear, and Ramesses III's pictorial evidence from Medinet Habu appears to indicate that an effective cohort held fewer than 10 men to a row. Furthermore, the Egyptian reliefs tend to show that the depth of such groups was greater than the breadth. Hence, we can place the dimensions more evenly as 10 men in breadth and 50 men in depth. This would give us 10 m width and 45 m in depth. This also seems unsatisfactory, especially as we must take into consideration that it is best, militarily or logistically speaking, not to advance with a lengthy horizontal force. One could assume that the company was divided in rows consisting of 20 men and in columns 25 men long, this being the minimum. Yet it may be better, at least considering the problematic visual representations, to make the columns longer. This I shall not contemplate here.

We now need ten such companies to make up the preferred total size of 5,000 men, and a 2–3 m gap fits the marching orders. (Note that I am not considering any chariotry at this point.) Here, we can assume that spaces existed between the men. Again, to place five companies in one large row would have extended the protective chariots forces too much and the division would have been harder to advance across terrain where the roads were of standard width. A division consisting of two columns or even one seems to have been the case. This way, the soldiers could move more freely on a Bronze Age road (never wide) and across difficult terrain, almost in single file, as befits the situations of hills and passes.

A further conundrum relates to the number of Egyptian troops which could get back to the camp of Ramesses after the sudden Hittite attack. I.e., did any flee and thus not participate in the assumed breakthrough of Muwatallis' chariots? We simply do not know anything related to this possibility, and thus it has been always assumed that none escaped or were able to flee northwards before the Hittites reached the camp of the Egyptians. It is still, nonetheless, worthwhile to calculate the (possible) flight of Ramesses' infantry in the second division. Here, we can argue with some degree of certainly that they would have, at the minimum, covered one km in six minutes. Hence, twelve minutes would have been used up. It is self-evident that the Hittite chariots could have outpaced any infantry that ran north but, of course, not any Egyptian chariot or chariots that may have decided likewise to take flight.

But how many enemy chariots were there? It is well-known that Ramesses explicitly ignores the size of the Egyptian infantry or chariotry sectors. He merely allows the poem to state that there were 2,500 Hittite chariots in the first wave and 1,000 later on.[43] This makes 3,500, if such integers can be trusted. However, for the sake of providing additional mathematical parameters, let us examine the length and width of these war machines. I will assume that those chariots ably discussed by Mary Littauer and Johan Crouwel provide a useful data source.[44] Here, the track or distance between the centers of the treads of both chariot wheels comes to ca. 1.70 m. I am relying upon the Tutankhamun chariots listed A1 to A 6 by Littauer and Crouwel. The axle length dimension, however, is the best to choose, and, on the average, we have 2.26 m in length. Spruytte, in his investigation, placed a total width for the Egyptian chariots weighing around 40–50 kg. This point he recently revaluated and, when analyzing the Florence chariot, noted that the total length of the axle was 1.985 m (with a track of 1.54 m). This information must be placed side-by-side with the effective width of one horse; i.e., the greatest thickness of the animal's barrel. Following Spruytte, we can note that with the ancient harness and a neck loop (and not coupled reigns or curricles), the two animals would not have "protruded to the left and the right of the extreme exterior or distal ends of the axle. In other words, let us accept the width of the Egyptian or Hittite chariot as around 2.25 m.

The distance between each chariot must be hypothesized. But there is the factor that chariots moving in marching order had to be arranged differently than when they were in battle. Following upon some comments of Colleen Darnell, would not an open, staggered arrangement in combat be more effective, considering the chariots as archer's platforms and not advancing in formation? For size, however, let me

43 The key references are P 84 with 132 (*KRI* II 31.6–9, 45.6–10) and P153 (*KRI* II 51.12–16).

44 Mary Littauer and Johan Crouwel, *Chariots and Related Equipment from the Tomb of Tutankhamun* (Oxford, 1985). They hypothesize different uses for these chariots, but their analysis might be queried at this point. Add Spruytte, "Le char antique du Musée Archéologique de Florence," in *Travaux de LAMPO* 1989, pp. 185–90, and Maria Christina Guidotti , ed., *Il carro e le armi del Museo egizio di Firenze* (Florence, 2002). Pier Roberto del Francia, "Il carro di Firenze," on pp. 16–37 of this booklet, provides the basic technological study of the Florence vehicle; Spruytte's research, however, is not mentioned. See as well Guiseppe Botti, "Il carro del sogno. (Per un grato ricordo personale)," *Aegyptus* 31 (1951): 192–98.

assume marching order. Here, perhaps a standard space of ca 2 m between chariots should suffice. For ten chariots we thus have: 2.25 × 10 + (9 × 2) = 22.5 m + 18 m = 40.5 m. For the length of a chariot we can follow, at an extreme, the pole length plus one half of the diameter of the wheel (including felloe thickness).

In this case the, pole reached the front part of the cab. Then we add the cab length from front to back with the additional factor of the wheel (.93/2 = .464). An approximation of 2.52 m — say 2.50 m — for the length of New Kingdom chariot poles is given by Littauer and Crouwel.[45] We must add the size of the back, which is ca. .468 m. In total we have ca. 3.45 m. I would allow more than a standard space of 1–1.25 m, following the lateral placement of the chariots, and set the space between rows as 2 m or so, perhaps even more, say ca. 2.25 m. To have a ten by ten or 100 chariot rectangle the length would come out to be: 10 × 3.45 + 9 × 2.25 = 34.5 + 20.25 = 54.75 m. In square meters, this is 1848. Thus the space occupied by 100 chariots in a symmetrical disposition was rather large. Given this, consider now 1,000 chariots of 100 × 100. The calculations are:

a) 2.25 × 100 + 99 × 1.25 = 225 + 100.25 = 325.25 m
b) 100 × 3.45 + 99 × 2.25 = 345 + 222.75 = 567.75 m

And in square meters we have 184,660.7.

Yet, we need not automatically disregard large and, especially, rounded figures. The assumed integer of 20,000 Egyptian men per division may fail as a near-precise estimate owing to these factors:

1) The time in marching as a unit. Here, one must hypothesize at least ten columns of men.
2) The presence of additional people connected with a marching army. Somewhat oddly or perhaps surprisingly, the camp followers connected with ancient armies, especially those of the Bronze Age, are never discussed.[46] How many of these plus non-combatants associated with a division (e.g., cooks, stable boys) can be assumed? How many delayed the advance of an army or, more importantly, how far behind the marching infantry and chariotry were these additional people? It is useful to press the latter point, as one may assume that the followers lagged after Ramesses' fourth division, that of Seth, and did not follow each division or even subsection of the total army.

45 *Chariots and Related Equipment from the Tomb of Tutankhamun*, pp. 79–80, 91. I am not considering either chariot A2 or that of Yuya and Tuyu.

46 On the other hand, Heagren discusses supply transportation in detail in his *The Art of War in Pharaonic Egypt*, pp. 194–207.

3) If 20,000 men per division is still argued, we still have to deduct the men who were only active warriors, and I also would assume that we must axe the army diarists as well as the artists, acting as modern photographers, who accompanied the troops. In fact, was it the case that these individuals belonged to the military, albeit in a non-combative role? If so, they ought to be considered to bear a status similar to military as physicians, whose purpose was most certainly not to kill one's opponent.
4) Finally, the natural attrition of life or, in this case, of war has to be entered into the equation. Some men were sick, others wounded or temporarily incapable of fighting. No platoon or regiment has ever been able to muster the full number of required men when the melee broke out. In fact, the results of the successful Hittite penetration of Ramesses' division two must imply that at least some Egyptian infantrymen survived. As we have seen, and this is not a new contribution on my part, the rapid deployment of Hittite chariots by Muwatallis was caused by his correct evaluation of the positional weakness of the pharaoh. So long as they could cut through Pre, then they would reach the Egyptian camp unopposed. Hence, time was of the essence and no delay was planned in order to mop up the slowly advancing troops in the division of Pre.

The battle around and within the Egyptian camp may have resembled at first a chariot encounter, where only Ramesses' elite warriors and himself parried off the Hittite chariot attack. It was not similar to the later campaigns of Merenptah and Ramesses III against the Libyans. One can surmise, perhaps with hesitation but nonetheless with some degree of certainty, that the fateful encounter of Hittites and Egyptians, the latter spearheaded by Ramesses, was similar to Thutmose III's encounter outside the city walls of Megiddo. Once more the elite subsectors of the ancient Near Eastern states of the Late Bronze Age determined the outcome. This was not, for example, the plan of Merenptah, who specifically states that his chariotry be supported by infantry whose ability as archers was paramount.

The points evaluated in this discussion remain *sub judice*. Few have been brought into discussion, owing to the presumed limited nature of the source material and its biases. Yet it is a requirement for analysis that logistical and structural aspects of any problem be addressed, and this is what I have done here. By and large, the battle of Kadesh has been discussed from many vantage points — military, religious, propagandistic, art historical, to name just a few. I have attempted here to set out the basic "sizes and shapes," so to speak, of the clash and to provide handy stepping-stones for future research. Therefore, the aim of this presentation is more directed at presenting a quantitative overview of the resulting melee than to reveal a hidden solution to the advance of the Hittites and their eventual repulse on day one.

# 5
# The Sinai of Seti I

This final chapter, although covering Egypt's imperialistic history in an era predating the great Ramesses II, nonetheless adheres to the main theme of this volume, namely, my emphasis upon time and the dating of events. As with the previous discussion, I shall concentrate upon warfare, but at the same time travel into partially charted oceans or, to be more historically accurate, into terrains both far removed and strange to the homeland of the New Kingdom Egyptians. Yet, with Ramesses's father, Seti I, we are fortunate that he too preferred a detailed and topographically accurate visual rendition of his military forays. Owing to this, I have purposely emphasized the iconographic nature of his pictorial record but not forgotten the historical background of his numerous campaigns.

In the Sinai campaign of pharaoh Seti I, a series of important iconic images emerges in full clarity for the first time during the New Kingdom.[1] This has a lot to do with the few Egyptian military reliefs dated to the Eighteenth Dynasty, as well as the completeness of the preserved Seti I material on the exterior northern wall of

1 At this point, I am referring to the Seti I war monument on the northern exterior wall of the Hypostyle Hall: Epigraphic Survey, *The Battle Reliefs of King Sety I, Reliefs and Inscriptions at Karnak IV* (Chicago, 1985). There is a handy diagram for the Seti I monument in *KRI* I 6. For this campaign, see Peter Brand, *The Monuments of Seti I: Epigraphic, Historical, and Art Historical Analysis* (Leiden and Boston: Brill; 2000), pp. 192–219, for Seti I's Karnak work; earlier, note William Murnane, *The Road to Kadesh: A Historical Interpretation of the Battle Reliefs of King Sety I at Karnak* (Chicago: Oriental Institute; 1985), pp. 55–64. For the problem of registers one and two on the east side, see my *War in Ancient Egypt: The New Kingdom* (Blackwell: Malden; 2005), pp. 188–202; and Clive Broadhurst, "An Artistic Interpretation of Sety I's War Reliefs," *JEA* 75 (1989): 229–34. The interpretation of Agnès Degrève, "La campagne asiatique de l'an 1 de Séthy Ier représentée sur le mur extérieur nord de la salle hypostyle du temple d'Amon à Karnak," *RdE* 57 (2006): 47–76, challenges the traditional or previously-held historical reconstructions of Seti I's wars. Her argumentation cannot be ignored, even though I find that I cannot follow all of her analyses.

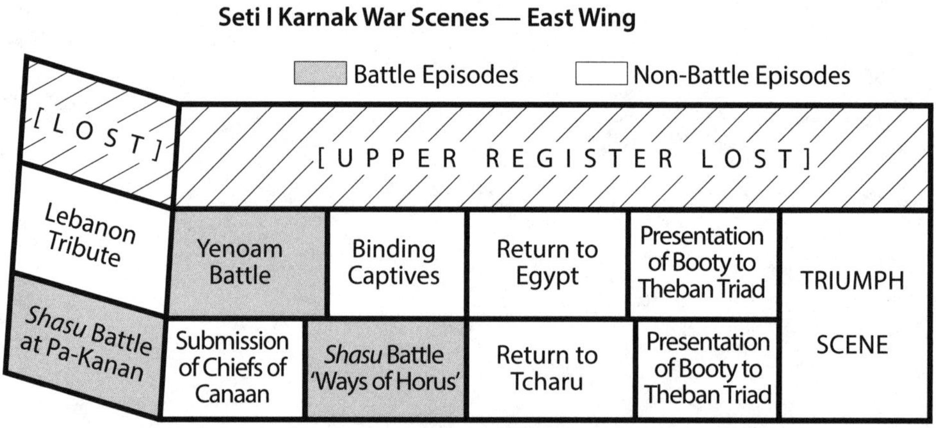


FIG. 8 Schematic Outline of the Seti I War Monument (diagram courtesy Peter Brand).

the Hypostyle Hall of Karnak. Nonetheless, as Nicholas Grimal has already observed, there are unusual or little-shown images that were copied onto this wall, ones that do not occur elsewhere.[2] In this discussion, I would like to highlight these icons and show how they work as the focus of the narrative of the king in the Sinai. First, however, let me describe the basic structure of the pictorial representations (fig. 8).

The Sinai campaign of this monarch is still, historically speaking, *sub judice*. Perhaps that is as it should be.[3] After all, Egyptologists and other scholars have been baffled by the seemingly unimportant military strength of the enemy. They are most

2 Nicholas Grimal, "La danse des peuples aux marches du royaume," *CRAIBL* (2001): 1159–83 for the basic study. Also useful is Stephen Quirke, "Frontier or Border? The Northeast Delta in Middle Kingdom Texts," in Alexandra Nibbi, ed., *The Archaeology, Geography and History of the Egyptian Delta in Pharaonic Times.* Discussions in Egyptology Special Number 1 (Oxford, 1989), pp. 261–79. Ellen Fowles Morris discusses the "neat, linear appearance" of the relief on pp. 407–8 of her important work, *The Architecture of Imperialism: Military Bases and the Evolution of Foreign Policy in Egypt's New Kingdom* (Leiden and Boston, 2005).

3 In this context, see these important studies: Antonio Loprieno, *La pensée et l'écriture: pour une analyse sémiotique de la culture égyptienne: quatre séminaires à l'École pratique des hautes études, Section des sciences religieuses, 15-27 mai 2000* (Paris, 2001), chapter 1; and Gerald Moers, *Fingierte Welten in der ägyptischen Literatur des 2. Jahrtausends v. Chr.: Grenzüberschreitung, Reisemotiv und Fiktionalität* (Leiden and Boston, 2001), chapter 3. Cf. Eliezer D. Oren and J. Shereshevski, "Military Architecture along the 'Ways of Horus' – Egyptian Reliefs and Archaeological Evidence," *Eretz Israel* 20 (1989): 8–22; with Oren, "The Establishment of Egyptian Imperial Administration on the 'Ways of Horus:' An Archaeological Perspective from North Sinai," in Ernst Czerny et al., eds., *Timelines: Studies in Honour of Manfred Bietak* II (Leuven, Paris, and Dudley, 2006), pp. 279–92.

FIG. 9 Seti I versus the Shasu (photograph courtesy Peter Brand).

definitely not an urban folk. Named Shasu by the Egyptians, these opponents of Seti I were depicted as very secondary protagonists.[4] Their lack of horses and chariots, the apparent rudimentary nature of their armor and war material, coupled with the absence of any chief directly opposing pharaoh in a "duel" situation — all point to the relatively inferior rank that they held in the eyes of the always triumphant and civilized Egyptians.[5] (See the depiction in fig. 9.)

The depictions of the Sinai campaign are contained only on the first register of the left hand or eastern side of the northern exterior wall. They proceed to a climax in typical Egyptian narrative fashion, and that scene will be found around the corner at the east. The king departs in his horse-drawn chariot,[6] meets the enemy in the field

4 For the Shasu, see the interesting but overly descriptive analysis of Thomas Staubli, *Das Image der Nomaden im alten Israel und in der Ikonographie seiner sesshaften Nachbarn* (Göttingen and Fribourg, 1991), pp. 35–58.

5 The following remarks are summarized from my *Icons of Power: A Strategy of Reinterpretation* (Prague, 2012). Note also Alice V. Baines, "The Function of Iconography as Autobiographical Narration in the tomb of Khnemhotep at Beni Hassan (Tomb 3), *JNSL* 21.2 (1995): 1–24.

6 The names of the horses (or of the teams ?) have been an interesting topic in Egyptological scholarship for the past few years. I can cite Aiman Eshmawy, "Names of Horses in Ancient Egypt," in Jean-Claude Goyon and Christine Cardin, eds., *Proceedings of the Ninth International Congress of Egyptologists* I (Leuven, Paris, and Dudley, 2007), pp. 665–76. But earlier, Laurence Caritoux, "Les chevaux de Pharaoh," *Egypt: Afrique et Orient* 11 (1998): 21–26, had covered this issue. This study was based upon her MA Thesis: *Le nom des attelages royaux au Nouvel Empire* (Montpellier, University of Montpellier, 1996). I must thank Laurence Caritoux for her kind assistance in sending to me a copy of her work. See now Pascal Vernus, "Réception linguistique et idéologique d'une nouvelle technologie: le cheval dans la civilisation pharao-

(and outside of a city or fortress),[7] prepares to return home with captives, and then presents his prisoners and booty to the chief god of the temple, Amun-Re. In essence, the pattern of snapshots, each of a separate or independent unity, presents the cinematic progression of the entire "film script." They allow one to stop and examine the key event. What we, the onlookers, see at first is the superhuman sized warrior-king, usually in his chariot, fighting alone in the pose of an archer against the foe.

One must commence reading these depictions in a direction away from one key zone, the doorway. This is due to the flow of the narrative, which begins in the direction outwards from the temple.[8] The doorway, or gate, forms the egress zone from which pharaoh departs. In the Seti I monument, the doorway on the north, flanked by three registers on the east and west sides, serves as the demarcation point. The king departs from his religious building and travels through foreign lands. Then he returns, proceeding in the route opposite to which he left Egypt. The two directions, in and out, are always carefully indicated.

Let us now analyze the first outbound scene. The king is shown fighting the Shasu Bedouin.[9] This scene is a typical one in structure and iconic representation and pres-

---

nique," in Myriam Wissa, ed., *The Knowledge Economy and Technological Capabilities: Egypt, the Near East and the Mediterranean 2nd millennium B.C.–1st millennium A.D.* (Barcelona, 2009), pp. 1–46.

7 See now James Hoffmeier and Stephen Moshier, "'A Highway out of Egypt:' The Main Road from Egypt to Canaan," in Frank Förster and Heiko Riemer, eds., *Desert Road Archaeology in Ancient Egypt and Beyond* (Cologne, 2013), pp. 485–510, a recent study concerning the "Sinai Corridor," an excellent term first used by Eliezer Oren to describe this administrative and military system.

8 Henry George Fischer, *L'écriture et l'art de l'Égypte ancienne: quatre leçons sur la paléographie et l'épigraphie pharaoniques* (Paris, 1986), p. 90 with fig. 20; and *The Orientation of Hieroglyphs: Part I. Reversals* (New York, 1977), pp. 41–47. See his comment in note 119 on p. 46: "the enemy is driven away from the entrance, on either side of it, and prisoners are lead to it." This is the arrangement of the historical war reliefs. Add Benoît Lurson, "La légitimation du pouvoir royal par l'observance des rites osiriens: Analyse d'une séquence de scènes de la grande salle hypostyle de Karnak," *Acta Orientalia Belgica* 15 (2001): 303–32; and "La conception du décor d'un temple au début du règne de Ramsès II: Analyse du deuxième registre de la moitié sud du mur ouest de la grande salle hypostyle de Karnak," *JEA* 91 (2005): 107–24.

9 Background information will be found in James Hoffmeier and Jacobus van Dijk, "New Light on the Amarna Period from North Sinai," *JEA* 96 (2010): 191–205. But the following are necessary to consult: See now James Hoffmeier, *Tell el-Borg I: Excavations in North Sinai* (Winona Lake, 2014); Anthony Spalinger, "Operational Bases: Gaza and Beth Shan," in Christina Karlshausen and Claude Obsomer, eds., *De la Nubie à Qadech/From Nubia to Qadesh. Études sur la guerre dans l'Égypte ancienne* (Brussels, 2016), pp. 63–80; Mohammed Abd el-Maksoud, *Tell Heboua (1981–1991)* (Paris, 1998), *Enquête archéologique sur la Deuxième Période Intermédiaire et le Nouvel Empire à*

ents the expected cut-off triangle of fleeing, maimed, or killed enemy strewn into a pile rising at the extreme east (fig. 9). But the importance of this depiction lies in the presence of various Egyptian fortresses and pools of water at the bottom of the scene. These military garrisons, first elucidated in a careful and painstaking study by Eliezer Oren and Joseph Shereshevski, are unique to these reliefs.[10] The reason for this is simple: Seti I had to traverse the Sinai in order to reach his next staging point at Gaza.[11] This, despite the hesitation of some scholars, must be the city depicted around the corner to the east: number three in the outwardly-focused snapshots.

As Oren and Shereshevski observed, the Egyptian fortresses are commonplace in presentation.[12] They look like small versions of a typical Asiatic city, whether it be

---

*l'extrémité orientale du Delta* (Paris, 1998), and "Tjarou, porte de l'Orient," in Charles Bonnet and Dominique Valbelle, eds., *Le Sinaï durant l'Antiquité et le Moyen Âge: 4,000 ans d'historie pour un désert* (Paris, 1998), pp. 61–65; Hoffmeier and Abd el-Maksoud, "A New Military Site on the 'Ways of Horus:' Tell el-Borg 1999–2001: A Preliminary Report," *JEA* 89 (2003): 169–97.

10 Oren and Shereshevski, "Military Architecture Along the 'Ways of Horus' – Egyptian Reliefs and Archaeological Evidence." Compare Spalinger, "The Battle of Kadesh: The Chariot Frieze at Abydos," *Ägypten und Levante* 13 (2003): 163–99; Oren, "The 'Ways of Horus' in North Sinai," in Anson F. Rainey, ed., *Egypt, Israel, Sinai. Archaeological and Historical Relationships in the Biblical Period* (Tel Aviv, 1987), pp. 69–119, "Ancient Military Roads between Egypt and Canaan," *Bulletin of the Anglo-Israel Archaeological Society* 10 (1982/1983): 20–24, "Migdol: A New Fortress on the Edge of the Eastern Nile Delta," *BASOR* 256 (1984): 7–44, "The Establishment of Egyptian Imperial Administration on the 'Ways of Horus:' An Archaeological Perspective from North Sinai," in Ernst Czerny et al., eds., *Timelines. Studies in Honour of Manfred Bietak* (Leuven, Paris, and Dudley, 2006), pp. 279–92. For earlier background I shall refer only to Abd el-Maksoud and Valbelle, "Tell Héboua II. Rapport sur le décor et l'épigraphie des éléments architectoniques découverts au cours des campagnes 2008-2009 dans la zone centrale du *khétem* de Tjarou," *RdE* 62 (2011): 1–17. But see the remarks of Brett Heagren in his *The Art of War in Pharaonic Egypt*.

11 On the military significance of Gaza, tactically speaking, see my study "Operational bases: Gaza and Beth Shan," in Christina Karlshausen and Claude Obsomer, eds., *De la Nubie à Qadech/From Nubia to Qadesh. Études sur la guerre dans l'Égypte ancienne* (Brussels, 2016), pp. 63–80. That analysis, written two years later than this one, is highly dependent upon my present remarks.

12 Add the useful remarks of Thomas Hikade, "Crossing the Frontier into the Desert: Egyptian Expeditions to the Sinai Peninsula," *AWE* 6 (2007): 1–22. He attempts to provide a "topos and mimesis" analysis to the accounts of various Sinai expeditions. He follows Loprieno by stressing the "completely foreign environment" at the northeast border of Egypt. The latter study is *Topos und Mimesis: Zum Ausländer in der ägyptischen Literatur* (Wiesbaden, 1988). Subsequently, note Thomas G. Gilroy, "Outlandish Outlanders: Foreigners and Caricature in Egyptian Art," *GM* 191 (2002): 35–52.

in Palestine or in Syria. Of course, they were something else. Indeed, as was proved, their models come from Egyptian conventions and represent transformations of a basic native pattern onto foreign cities. Hence, it is not surprising that all of the urban structures in the New Kingdom war reliefs look essentially the same. They have massive walls, tall gates of wood, crenellated battlements, and sometimes, in order to show their importance, a second "crenellated turret" in the middle. Thus, as was demonstrated, their nature is iconic and does not reflect the actual type of buildings of city walls that were encountered during Egyptian forays into Western Asia.[13]

The wells or cisterns plus this series of fortresses indicate the degree of Egyptian control over the coastal road from the east Delta through the Sinai.[14] The Shasu Bedouin, not associated with any of these fortifications, remain always outside of the garrisons. In addition, we see no Egyptian troops in any of these military posts. The major urban center depicted in the king's Karnak monument appears to be completely uninhabited. The same can be said with respect to the smaller Sinai fortresses, the latter being of lesser importance. Is this to be expected? In order to proceed with this analysis let us turn to the method of depiction.

In the first battle scene, the three fortifications and the three accompanying wells are located in a small but significant portion of the bottom section of the relief.[15]

---

13 Concerning Nubia in this context, see now Robert Rittner, "Libyan vs. Nubian as the Ideal Egyptian," in Stephen E. Thompson and Peter der Manuelian, eds., *Egypt and Beyond: Essays Presented to Leonard H. Lesko upon his Retirement from the Wilbour Chair of Egyptology at Brown University, June 2005* (Providence, 2008), pp. 305–14. He gives a short analysis of an interesting dichotomy. The two groups at this time (Dynasty XXV) were no longer the antithesis of sedentary/civilized. Thomas von der Way, in his *Göttergericht und 'Heiliger' Krieg im Alten Ägypten: Die Inschriften des Merenptah zum Libyerkrieg des Jahres 5* (Heidelberg, 1992), pp. 25–28, also discusses alterity in the Ramesside Period. Concerning the contemporary Egyptian view of Libyans, Asiatics, and Nubians, cf. Roland Tefnin, "La perception de la différence en Égypte pharaonique," *Civilisations* 35 (1986): 39–55.

14 Most important in this context is the study of Morris, *The Architecture of Imperialism* (Leiden and Boston, 2005) pp. 344–59, 402–47. Valbelle's beginning contribution was "Entre l'Égypte et la Palestine, Tell el-Herr," *BSFE* 109 (1987): 24–38. The subject is voluminous, but Oren and Shereshevski's article cited in note 3, above, remains basic. See as well Paul Figueras, *From Gaza to Pelusium: Materials for the Historical Geography of North Sinai and Southwestern Palestine (332 BCE–640 CE)* (Beer Sheva, 2000), passim, especially chapters 3 and 4. Pp. 122–23 are particularly devoted to Mount Casius.

15 Connected with the Seti I data is the question of the route of the Exodus and the location of the Biblical Migdol. As the issues remain *sub judice* and the secondary literature is voluminous, let me cite only a few recent studies: James K. Hoffmeier in *Ancient Israel in the Sinai: The Evidence for the Authenticity of the Wilderness Tradition* (New York, 2005). On pp. 95–96 he discusses these geographic conundrums, and his tentative conclusions are presented on p. 96: the excavations of the past twenty years at Tell

There can be little doubt that they were added *after* the plan of the depiction had been approved. In order to examine the artistic set-up, it is now necessary to describe the scene in its totality.

The standard method operated, one might say, as always.[16] First, the cartoon or template of the king was transferred to the wall. Seti I is presented on order, so to speak, shooting his arrow downwards into a mass of overwhelmed Shasu. He is alone in the cab of his chariot and his horses arch upwards. This image must have been the first portion to be designed (if not carved), owing to its singular importance: king first, verdict afterwards. Moreover, from later aspects of the original positioning of this icon (king, chariot, and horses) in the reliefs of Ramesses II, it can be argued that the crucial nature of the royal depiction preceded all other images to be included within the designated snapshot. In subsequent depictions of king versus enemy citadel, the enemy is encountered fighting on foot and sometimes in chariot before the king. Here, the weak enemy shows no resistance. The plot is developed by means of a few spotlighted scenes, as one learned to do with respect to modern novel writing.[17]

---

Qedua and Tell el-Herr make it clear that the former could be the Migdol of the biblical prophets and perhaps the place with the same name mentioned in Esarhaddon's annals of his invasion of Egypt, but not Magdalos or Magdolum of the Classical period. On the other hand, Tell el-Herr, six km south of Qedua, seems to fit with the location of the later site, as it remains the only late Persian to Greco-Roman Period fort between Pelusium (Tell Farama) and Sile (Tell Abu-Sefêh). Finally, the archaeological work at these two tells also eliminates them from being the thirteenth-century BC fortress, the Migdol of Menmaatre. Following Hoffmeier: "As Kenneth Kitchen observes, The New Kingdom 'Migdol' of Sethos I is identical with neither of these sites but remains to be discovered somewhere in the vicinity." Oren's "Migdol: A New Fortress on the Edge of the Eastern Nile Delta," *BASOR* 256 (1984): 7–44, is nonetheless crucial. In this analysis, his archaeological observation was that the XIXth Dynasty Migdol has to be searched elsewhere than at Qedua and Tell el-Herr. He was the first to prove this.

16 As background data let me cite the following: Roland Tefnin, "Élements pour une sémiologie de l'image égyptienne," *CdE* 66 (1991): 60–88, and his "Image, écriture, récit. À propos des représentations de la bataillle de Qadesh," *GM* 47 (1981): 55–73. For later studies, see Éric van Essche-Merchez, "Pour une lecture 'stratigraphique' des parois du temple de Ramsès III à Médinet Habou," *RdE* 45 (1994): 87–116, "Quelques réflexions sur l'espace et le récit à Médinet Habou," *Annales d'Histoire de l'Art et Archéologie* 11 (1989): 7–24, and "Scène, paroi, récit. Des cases de bande dessinée aux reliefs ramessides: pour une complémentarité méthodologique," *Art & Fact* 13 (1994): 40–57; and subsequently, David O'Connor, "The Sea Peoples and the Egyptian Sources," in Eliezer D. Oren, ed., *The Sea Peoples and their World: A Reassessment* (Philadelphia 2000), pp. 85–102.

17 Let me refer to the helpful remarks of Francis Otto Matthiessen, *American Renaissance: Art and Expression in the Age of Emerson and Whitman* (Oxford, 1940), p. 203.

The movement of the chariot forward or away from the doorway is related to the upward direction of the horses.[18] They purposely form an acute angle with the horizontal so that the chaos of the king's opponents can fill the empty spaces above and below the horses' bellies, their front legs, and heads. It is a norm that vertically arranged inscriptions of a rather banal nature might be included above and to the front of the horses.[19] Nonetheless, in that case the enemy would be situated more to the front of the chariot team and thus the image would tend to be more elongated in order that the entire enemy is depicted.

The positions of the above two portions of a typical battle were, in this case, supplemented by the additional presence of the three fortresses and three wells. They could only be placed where space was still available, under and to the left of the horses' bellies and front legs. In addition, these features were simply conjoined to the whole. Whether or not Seti I actually met the Shasu in the vicinity of six fortresses must remain an open question. (Note that there are three additional fortresses and wells included in Seti's battlefield presentation snapshot.) Perhaps the course of battle involved a movement northwards and slightly westwards (?) from the row of Egyptian garrisons. But with no difficulty involved, these representations were artistically placed one after the other. This means, of course, that the third depiction around the corner to the east, logically the last, may designate the place where the Shasu finally resisted the Egyptian ruler. (On this problem, see our comments below.)

Nested symmetrically between two identical scenes of war is the non-combative image wherein Seti I is preparing to leave the battlefield and has already directed his chariot homewards. This spotlighted event forms a final visual reflection of the early Sinai warfare of Seti I. We move from the first depiction, king in battle, to the victorious outcome.[20] The latter relief was purposely assigned space at the corner, so that by this means an obvious completion of physical martial activity was indicated. Seti I fights with his enemy in the first relief and then is ready to depart. His return — or at least his departure from the battlefield — was logically positioned away from the doorway and separate from the eventual return of the rest of the Egyptians to Egypt. But it was necessary to present the pharaoh facing the doorway and not moving outwards from home or, in this case, to the east. (See below for a more complete analysis of this situation). It is for these reasons that a simple reading

18 The basic positioning of horses, troops, pharaoh, etc. is ably discussed by Susanne Constanze Heinz, *Die Feldzugsdarstellungen des Neuen Reiches: Eine Bildanalyse* (Vienna, 2001). The crucial reliefs are reproduced on pp. 242–46.

19 See Pascal Vernus, "Les espaces de l'écrit dans l'Égypte pharaonique," *BSFE* 119 (1990): 39–43.

20 The image of the "King in Battle" is, of course, part of the ruler's warrior ethos. It is a good example of what Erik Hornung called "Pharao Ludens," *Eranos Jahrbuch* 51 (1982): 479–516; add my "Ramesses Ludens et Alii," in Panagiotis Kousoulis, ed., *Studies on the Ancient Egyptian Culture & Foreign Relations* (Rhodes, 2007), pp. 71–86.

of a progression from A (battlefield encounter) to B (booty departure) to C (battle near Pa-Canaan/Gaza?) cannot hold.[21] Seti I should be going further *away* from his country. Hence, he ought to depart from the first battle northwards or at least outwardly instead of inwardly.

The artistic convention of the battlefield presentation snapshot, on the other hand, required that the king be ready to leave the area, either just getting into the chariot or looking to the rear towards his soldiers and captives. His horses, however, would be facing home. Thus, the direction of pharaoh is reversed at this point if we consider "truth" to be a main component of this Egyptian artistic representation.[22] If drama, instead, is involved, then the baldly empirical approach of interpretation risks contradiction.[23]

Most definitely, if Seti I moved farther north, then his rightward positioning in the direction of the doorway makes no sense. We thus have to view the second depiction (B) as completing the narrative of the first image.[24] It is one that is self-standing and needs no further elaboration, temporally or geographically,[25] except for, of course, the final image of a return home with or without an additional presentation scene, topos of receiving booty. What ought to happen next is the return to Egypt. Hence, the direction of Seti I is directed homeward even though the king actually proceeded farther northwards.

---

21 I cannot see any other solution to the sometimes perplexing equation (and to some people only) of Pa-Canaan = Gaza but to accept it. Note the comments of Morris, *The Architecture of Imperialism*, p. 443.

22 Cf. G. A. Gaballa, *Narrative in Egyptian Art* (Mainz, 1976) retains its usefulness in this context. The review of this work by Claude Vandersleyen, *CdE* 53 (1978): 275–78 effectively counterbalances Gaballa's noteworthy empirical analysis.

23 Hence, see my "Re-Reading Egyptian Military Reliefs," in Mark Collier and Steven Snape, eds., *Ramesside Studies in Honour of K.A. Kitchen* (Bolton, 2011), pp. 475–91. I part somewhat with the chapter of Hoffmeier, "The Gate of the Ramesside Fort at Tell-Bourg, North Sinai," pp. 207–18, in the same volume.

24 In other words, each register section was self-contained. Such an interpretation is significant, because it would then appear that each register refers to a separate campaign. On the other hand, this is not necessarily the case. As I have stressed in the study referred to in the previous note, dramatic considerations as well as artistic sensibilities must be taken into consideration. I would, therefore, put the interpretation into a different perspective: each register completes a thought or narrative process. Whether or not it belonged to the preceding or the following cannot be determined by mere circumstance.

25 Cf. Michelle I. Marcus, "Geography as an Organizing Principle in the Imperial Art of Shalmaneser III," *Iraq* 49 (1987): 77–90. I find her later study, "Geography as Visual Ideology: Landscape, Knowledge, and Power in Neo-Assyrian Art," in Mario Liverani, ed., *Neo-Assyrian Geography* (Rome, 1995), pp. 193–202, somewhat overdone.

Thus the second battle between Seti I and his (Bedouin) opponents was easily assigned a location separate from the preceding two scenes. It could be placed "around the corner," so to speak. One could, therefore, avoid seeing that event by remaining directly in front of the northern exterior wall, while still understanding the pharaoh's entire war in the Sinai. If omitted by a viewer, the following arrangement would have been available. (Later, I shall return to snapshot C.)

A: Battle with Shasu
B: Preparation for departure after victory
C: Return to Egypt (see below)
D: Presentation of captives and booty to Amun.

The physically stretched or widened pictorial relief of D involved certain preliminary structural arrangements. First, an overt demarcation was chosen, and it was organized as a vertical bar. A moderately wide barrier of water was required, against which the master artist located the "gateway" into Egypt. Following the direction of activity, the return homewards, that gateway was positioned to the west (visually, to the right). Next to it and forming a symmetrical design with the vertical water channel was a relatively wide column of hieroglyphs. To the extreme right there are two registers of greeting Egyptians, the upper and lower groups clearly distinguished from each other. Note that both are called "priests and chiefs of Upper and Lower Egypt." If there is a hierarchy between the two, it can be seen in the performances and acts, as well as the clothes of both: the lower group is more significant, whereas the upper cohort makes the background music.

But we must focus upon the left > right portion of this wide scene (fig. 10). There, Seti I is depicted in the typical victory mode of arriving at the literal gate of Egypt with his prisoners, also in two registers, behind the royal chariot. Between the pharaoh and the other humans, small captions were later added, all of which are quite short and simple. Male prisoners precede the monarch, in three groups, from top to bottom, of 9, 5, and 7. The last final three men just look behind the chariot. Perhaps an obvious simple arithmetical progress is apparent (9, 5 —down 4 —, and 7 —up 2; the average is 7).

The internal positioning in this vignette is also overtly artificial. There is one local Sinai fortress, conveniently located under the chest of the two horses that are propelling Seti I to the right and homeward. This is, in fact, the only space where any garrison could be depicted. To the front is the lowest register of prisoners, and then even more to the fore in the middle register is a central gateway. To the rear are two separate registers: the three captives above and the king's son below. All possible locations for a Sinai fortress were already taken, save the rather uninspiring or secondary status of the free space under the foreparts of the royal horse team. Finally, the stylization of registers can be recognized. From rear to front — from foreign

Fig. 10 Seti I at Pa-Canaan (photograph courtesy Peter Brand).

clime to native land—we commence with two subsidiary registers. In front of Seti I there are three registers of prisoners in, antithetically, a final bipartite division at the extreme right. The order is therefore: 2 > 3 > 2, with the demarcations of king and vertical water channel forming the breaks.

Two additional local Egyptian garrisons, both linked with their cisterns or wells, may be found at the expected bottom of this final narrative depiction. Note that they, too, have been added into the scene at the last moment. One is placed directly under and in the center of the ventral portion of the two tethered horses, whereas a second is located behind the rear legs and under the tail. These two images also contain the expected wells or cisterns. Owing to the limitation of space, not every well can be seen, but at least four are indicated by the small accompanying texts.

Should we assume that it was at this point that Seti I gathered together his troops for the return to Egypt? Once more I query the assumption of the verisimilitude of these representations. The latter segment is pleasantly juxtaposed with another two-register system in the final scene. There, Seti I presents a four-level or four-register parade of booty, followed by two registers of prisoners: ten on the top, and on the bottom probably nine or ten men. Yet, at this point the artificiality of the visual presentation is completely transparent. This factor, that of realism versus verisimilitude, must be continually examined in the scholarly literature in order to deflect a hyper-reliance upon the veracity of Egyptian narrative wall reliefs.[26] Nowhere else than in

26 This aspect of narrative relief work is discussed in my final chapter (17) of *Icons of Power*, "Summary and Possibilities for Further Investigation." Additionally, let me cite this useful study: Jan Assmann, "Flachbildkunst des Neuen Reichs," in Claude Vandersleyen, ed., *Das alte Ägypten* (Berlin, 1975), pp. 304–25.

the final presentation scene do we see the overt effects of cultural interpretations at work. Alterity, at the minimum, must be considered, and how demanding such a schema was can only be stated after a detailed study of iconic representation is done.[27]

Although it is partly outside of the scope of this study, I cannot leave the scenes of Amun and Seti I without mentioning a few salient characteristics of the lack of "truth" that occurs in all pictorial narratives of war. First, the statue of Amun is as large as the pharaoh. This was usually not the case. Second, there are no accompanying priests in the depiction. They are precluded, owing to the intimate relationship in which father Amun and son Seti I appear. The father-son topos or "constellation," as Jan Assmann has called it, is nowhere better viewed than in these final snapshots.

We never see Amun during battle, and this is to be expected. He is absent as well when the king marches out of Egypt and when he returns home. Indeed, he never appears in momentous events, such as that of Ramesses II at Kadesh.[28] There, it is true, the pharaoh implores his father, but the deity is never present when events occur, although I grant the presence of the ram-headed standard of Amun in the Medinet Habu reliefs. In fact, the history visually presented is one wherein only mortals act. No hidden hand emerging from a cloud or a figure from on all high is ever allowed to intrude into the virile deeds of pharaoh. He alone acts. His godhead is, at the minimum, passive throughout the entire voyage thither and return hither. The reason for this approach is directly connected with the involvement of the god-father and his son, the Egyptian ruler. To be sure, the latter is given the prognostication of success at home, in the Karnak temple, to which the ruler is obliged to return after campaigning. Hence, we never see the departure of the Egyptian monarch to war within a religious setting. As we see him for the first time on the temple walls, he is already moving away from home. Either by mustering troops or traveling to a foreign

---

27 The attitude of some recent Graeco-Roman scholarship on the matter of alterity is contained in Beth Cohen, ed., *Not the Classical Ideal: Athens and the Construction of the Other in Greek Art* (Leiden, 2000). The volume of Jean Riaud, ed., *Étrangers et exclus dans le monde biblique. Colloque international à l'Université Catholique de l'Ouest. Angers, les 20 et 21 février 2002.* Théolarge 3 (Angers, 2003), can be mentioned in this context. An overview of alterity in the Hebrew, Akkadian, and Canaanite languages, based on the Amarna Letters and specifically versus Egyptian attitudes, may be read in Mario Liverani, "Stereotipi della lingua 'altra' nell'Asia anteriore antica," *VO* 3 (1980): 15–31. His "The Ideology of the Assyrian Empire," in Mogens Trolle Larsen, ed., *Power and Propaganda: A Symposium on Ancient Empires* (Copenhagen, 1979), pp. 297–317, presents a very worthwhile study similar to our approach. For general remarks, add *Prestige and Interest: International Relations in the Near East ca. 1600–1100 B.C.* (Padova, 1990).

28 This is an important point subtly deflected by Jan Assmann in "Krieg und Frieden im alten Ägypten. Ramses II. und die Schlacht bei Kadesh," *Mannheimer Forum* 1983/84 (1984): 175–231.

land, the Egyptian king is placed physically outside of the doorway or entrance of any temple. Therefore, in a true sense he has already departed from that building.

Seti's second military confrontation in the same bottom register is, as already observed, purposely located around the corner of the exterior northern wall of the Hypostyle Hall. This event is self-contained. One can read it without viewing any of the previous two visual snapshots. Here, the representation is a standard one: king in chariot aiming his arrow against a mass of enemies who are located, *de rigueur*, in front of him. The visual angle of trajectory moves, as normal, in an acute, almost 45-degree angle upwards. There is no citadel in which the defeated foes remain under siege. After all, are these enemies not uncivilized, failing to partake in any effective social or cultural norm, such as the city states of Palestine or Syria? Yet the town of Pa-Canaan, normally identified with Gaza, is attested, performing its iconic role as the cities of Asia always do. With the surrounding moat and obligatory tree, the whole scene appears at first to be stereotypical, merely offering the commonplace geographic topos of king in battle at an Asiatic city.

What is significant in this depiction, nevertheless, is the presence of detailed signs of despair, mourning, and total abnegation. The artistic methods chosen to convey these severe, or at least sharply accented, moods of resignation can only be found elsewhere in Ramesses II's Beit el Wali Nubian reliefs. Both scenes make the enemy completely ephemeral.[29] Their resistance was useless. They cannot even mount an effective challenge against pharaoh. No major opponent can be found leading the resistance. Whether the king is Seti I or Ramesses II, the enemy is rendered totally inept without any show of defiance. Indeed, both series of wall reliefs feature women consoling their tribespeople or else children being comforted if not led away from the carnage.[30] In the Pa-Canaan/Gaza scene, there are parallel reactions to the ruler's success: suing for peace, ministration to wounded or exhausted Bedouin by their own kind, and the removal of at least one child from the melée.

---

29 Cf. *Icons of Power*, chapter 3, for a classic icon (monkeys), one that is overtly reflective of the Egyptians' low regard for the Nubians. See Herbert Ricke, George R. Hughes, and Edward Wente, *The Beit el-Wali Temple of Ramesses II*, (Chicago, 1967), pl. 9. For the iconic role of monkeys in similar depictions, see R. D. Barnett, "Monkey Business," *JANES* 5 (1973): 1–5.

30 The use of the so-called "genre" scenes, obviously employed to accentuate the total collapse of the enemy's morale, reminds one of the remarks of Matthiessen referred to earlier in note 17, above. He notes Nathaniel Hawthorne's use of "oddities of low comedy," a practice that can be traced at least back to Shakespeare (cf. especially the evening before the battle in *Henry V*). These representations, such as the presence of a drunken or sleepy watchman, the hurt leg of a soldier in the Abu Simbel Kadesh reliefs, or physical disputes among the soldiers heighten the experience of ongoing quotidian activity. None are real; all provide a visual backdrop to the main action of the spotlighted event.

This juxtaposition of varying reactions to the Egyptian success is to be set in contrast with Ramesses II's overwhelming campaign against the Nubians. At Beit el Wali, however, the narrative flow is straightforward. The king attacks from the right and the enemy proceeds in hasty flight to the left. We move out and away from pharaoh to the acute angle of enemy soldiers until they merge into a series of running infants. The viewer is thus directed more and more to the left until some men are shown arriving at the homestead at which a more domestic scene predominates, punctuated, of course, by the news of the defeat.

Great tableaus were particularly used in the decorative relief work of the XIXth Dynasty. Hence, by the reign of Seti I, pictures contain a great deal of drama and excitement.[31] I believe that this analysis still holds true, despite the subsequent publication of Dynasty XVIII narrative military scenes. Most certainly, Assmann's rejection of the common icon of the king in battle representing merely the triumph of Maat over Chaos deserves our attention.[32] His interpretation added to these two diametrically opposing forces those of fear and trepidation ("Angst und Schrecken"). He concluded that "the drama of the moment" was an innovation of Dynasty XIX. Indeed, despite the later published evidence of Ahmose, Amunhotep II, Tutankhamun, and Horemheb (Saqqara tomb), it remains the case that no clear-cut evidence of such a "dramatic turn" in relief depiction of historical war scenes can be argued for any XVIIIth Dynasty relief.

Not having so much space on which to imprint all of these events, the Egyptian master designer at Karnak allowed for some type of caesura at the inner corner and gave the expected reverse flow to the action. Here, the flight of the semi- human barbarians is represented not by a line of running soldiers but rather a chaotic mass in which three small but significant and independent scenes can be discerned. That group becomes divided into three registers, with a fourth located closer to Seti I and composed of the peace negotiators. And because the city-fortress plays no direct role in the pictorial narrative, it remains seemingly uninhabited, outside of the carnage. The differences between this scene and the other one where Seti I also fights Bedouin are as follows:

---

31 Assmann, "Flachbildkunst des Neuen Reichs," p. 312. John Baines adheres to the opinion that the military reliefs of Seti I "constitute an important exception" to his thesis that this "art was not strongly innovative": "On the Status and Purposes of Ancient Egyptian Art," *CAJ* 4 (1994): 85.

32 In this context one can refer to Tefnin's work, "Discours et iconicité dans l'art égyptien," *GM* 79 (1984): 55–71. Tom Hare, *Remembering Osiris: Number, Gender and the Word in Ancient Egyptian Representational Systems* (Stanford, 1999) appears to be in debt to Tefnin. See his comments on p. 88. Yet it was Henry Fischer who, above all, was the first to stress and elucidate in detail the relationship between image and writing in ancient Egypt.

| I: Along the Sile to Gaza Road | II: At Pa-Canaan |
| --- | --- |
| fortresses and wells present | topos of city hill (tell) present |
| no peace initiative | "elders" (?) implore Seti I for peace |
| enemy flee | enemy flee but specific actions are present |
| no weapons present | enemy have spears |
| enemy have no leather armor on chest | enemy have some type of armor on chest |
| enemy have elongated rounded caps | caps (for protection) are rounded |

Clearly, there are differences between the two opponents whom Seti I encountered on the Sinai road. The Egyptian artists or designers of reliefs made a good attempt to represent the various ethnic and social groups who lived outside of their country and presented their different human images in a deliberately exacting fashion.[33] Dress and color of skin were distinctive, as was physiognomy. Beards for Semites, kinky hair for Nubians, light skin for Libyans (and the ubiquitous feather and penis sheath) are but three of the multifarious differing characteristics that were employed. Of equal importance were the tools of war. Granted that slips in representation occurred, quite often to the amusement of modern Egyptologists — witness the two Hittite men to a chariot in certain scenes of the battle of Kadesh or the Libyans intruding in an Asiatic war scene — the low level of war preparedness as represented in the Nubian and Shasu depictions cannot be left without comment.

Certain key attributes of a worthy opponent were necessary according to the empire artists of pharaonic Egypt. I have already remarked upon the lack of sophisticated body armor in both of these scenes. We can add the absence of any chief opponent. Seti I need not engage in a "duel," which he would win, of course. All that is necessary is to crush the enemy either under the wheels of his chariot or by means of his arrows. I am more than willing to recognize the lack of one hundred percent accuracy in the depiction of foreigners, especially when the Egyptians were faced with new peoples or a variegated and motley group of soldiers. Witness, as John Darnell indicated, the difficulties in identifying some military opponents as northerners and Hittites. Yet there is ample evidence to conclude that, *mutandis mutatis*, these depictions of warfare attempted realistic image making.

33 John Coleman Darnell, "Supposed Depictions of Hittites in the Amarna Period," *SAK* 18 (1991): 113–40; and earlier William Ward, "The Shasu 'Bedouin:' Notes on a Recent Publication," *JESHO* 15 (1972): 35–60. The subsequent analysis of Ann-Laure Oosthoek, "Hittite ou pas Hittite? Trois représentations à caractère hybride," in Claude Obsomer and Ann-Laure Oosthoek, eds., *Amosiadès: Mélanges offerts au professeur Claude Vandersleyen par ses anciens étudiants* (Louvain-la-Neuve, 1992), pp. 335–46.

William Murnane, in his two editions of *The Road to Kadesh*, his introduction to the battle reliefs of Seti I, and his later study on Egyptian kingship was notably anxious over his interpretations of the Shasu-Sinai war of king Seti I.[34] His indecisions concerning the chronological arrangement of the pharaoh's wars are intricately woven, but his conclusions remained always tentative. For our purposes, it was the problems associated with Seti I's opponents that troubled him greatly and concerning which we have professed some, perhaps hesitant and tentative, explanations. It is sufficient to point out, however, that more original indecision and confusion can be noted in the final scenes of Seti I presenting his captives and booty to the god or gods of Karnak. The presentation snapshot in the first register on the east side, for example, contains two subregisters of Asiatics, none of whom can be clearly identified with the Shasu. Once more, we are facing the same difficulties that Frank Yurco had with respect to his presumed Israelites in the scenes of Merenptah's Canaanite war: how accurate are Seti I's portrayals of Shasu versus "Israelites"?[35]

These of Seti I are not. Bound captives, standing behind the victor, do not have the two bands across the torso so clearly seen on the king's opponents in the key Shasu attack. They also lack the leather armor worn outside of the city of Pa-Canaan. Indeed, there are some captives who wear thin garments above the waist that cannot be identified in the battle scenes. On the other hand, in the return of Seti I to Egypt, his prisoners wear different caps; at least two types can be recognized. It is also true that the first man in the top register of this scene appears to be a leader or a chief of the enemy. This might be surmised on the basis of his garment, which is fuller than any other that the Asiatics in register one possess. In addition, the presence of some type of shawl is also to be stressed. Note that the other captive northerners wear either the tasseled caps of the Shasu depiction or the more rounded and

---

34 His equivocations with regard to presenting a coherent historical analysis may be seen in the attempt to define the role of the "mysterious" Mehy: "The Kingship of the Nineteenth Dynasty: A Study in the Resilience of an Institution," in David O'Connor and David Silverman, eds., *Ancient Egyptian Kingship* (Leiden, New York, and Cologne, 1995), 199–200. Compare this reconstruction to that presented earlier in his *The Road to Kadesh*, pp. 163–75.

35 Cf. Spalinger, *Icons of Power*, chapter 5. The background study is Frank Yurco, "Merenptah's Canaanite Campaign," *JARCE* 23 (1986): 189–215. There have been numerous follow-up studies dealing with his important discovery, but it is best to supplement Yurco's original work with that of "Merenptah's Canaanite Campaign and Israel's Origins," in Ernest S. Frerichs and Leonard H. Lesko, eds., *Exodus: The Egyptian Evidence* (Winona Lake, 1997), pp. 27–55. Brand has returned to these scenes in the useful study "The Date of the War Scenes on the South Wall of the Great Hypostyle Hall and the West Wall of the *Cour de la Cachette* at Karnak and the History of the Late Nineteenth Dynasty," in Collier and Snape, eds., *Ramesside Studies in Honour of K. A. Kitchen*, pp. 51–84.

curved caps. In the snapshot of the attack outside of Pa-Canaan there are no tassels on any headdress. One example, possibly treacherous as it might be misunderstood, shows a defeated enemy without any headgear. (He is situated under the forepart of the torso of the king's horse team.) Yet it is possible that some of the discrepancies would have been erased with the painting of the reliefs.

It is unnerving that the final scenes of ritual presentation differ somewhat with respect to the garments on the upper torsos of the enemies. Indeed, here as well, one might wish to view the men closest to the king as leaders of some sort. Most certainly, they all lack those caps with tassels. Therefore, I think it wise not to be completely dependent upon the veracity of representation and to follow a more cautious approach, such as that offered by Darnell or even Murnane. Yet even if we shy away from a definitive conclusion, note that in register two the captives displayed to the Theban triad of Amun, Mut, and Chonsu are dressed differently from those below. In particular, see the full-length garment and its tied knot set at the middle of the lower torso. Moreover, the beards of the men are fuller than those of the Shasu and their associates, and not at all pointed. We can thus easily differentiate the urban dwellers of the second register on the west side — who lived in the city of Yenoam and its allies — from the marauders of the Sinai. If only because of this clear differential in dress, one may surmise that the king's campaign in his first regnal year reached Gaza.

The visual icons of war relating to the Sinai campaign can be summarized as follows.

1. Presence of fortresses with cisterns or wells. They form the typical "foreign" and northern orientation of the depictions and are included only in three snapshots:
    a. Fight on battlefield, a scene that serves as the "departure" snapshot
    b. The battlefield presentation
    c. The return home.

   In one case (b) there is a detailed Sinai fortress which is well-placed above the backs of the king's horses. This is a noteworthy divergence from the expected location at the very bottom and, significantly, the garrison is a more impressive one. Nevertheless, observe that it is virtually identical to the commonplace Egyptian depictions of foreign cities located on tells. This example, in fact, supports the earlier analysis of Oren and Shereshevski to a tee.[36]
2. The city of Pa-Canaan is correctly placed on a tell. Therefore, it is not a mere garrison. The icon performs the role of a marker of finality, the last place to which Seti I went. Not surprisingly, we have advanced farther north and therefore no cisterns or fortresses are to be seen.

36 "Military Architecture along the 'Ways of Horus.'"

3. This city scene is a commonplace one in the urban setting of Western Asia. But the presence of the enemy outside of Pa-Canaan is indicative of the non-urban nature of the enemy. Moreover, the depiction includes minor yet helpful visual indications of the enemy's despair and complete desolation owing to the Egyptian victory. Because this is never presented in the common topos of an Asiatic defeat, this is a further reason why we must separate the Pa-Canaan event from the expected representations of Asiatic warfare.
4. Horse names. Suffice it to say that this same depiction gives proof that the names of the king's steeds referred only to one of the two animals: the visible front horse. The situation is seen most clearly in the presence of two names within one snapshot. Seti's right hand horse is identified below and the upper one to his left. Hence, as others have noted, *ḥtr* in this case means "horse."[37]
5. The clothing (or dress) but also beards and hair differentiate Seti's register one opponents from the "normal" Canaanites or Syrians whom the Egyptian pharaohs met while advancing northwards into Western Asia.

Thus, the painterly aspects of Seti I's Karnak reliefs emerge in a clearer light than if one practices the straightforward historical-artistic interpretation.[38] As a result, the demonstrative nature of the final visual result can be appreciated better, owing to an understanding of its artificiality. The highlighting of separate events with a chief protagonist was nothing original. That means of narrative presentation was age-old. New, however, was the attempt to provide a visual flow in which a climax could be reached and in which subsidiary aspects, none of which were directly concerned with

37 See the references given above in notes 3 and 10.

38 Assmann, "Flachbildkunst des Neuen Reichs," also referred to the filling of background space in such relief work wherein one can now find a plethora of images, such as trees or houses, which present a needed localization of the event. Yet one can also note the "eradication" of the "non-space" through columns or lines of hieroglyphs. Following James Henry Breasted's brief comments, two independent eye movements were expected, one down and the other across: *The Oriental Institute*[2] (Chicago, 1933), p. 223. These military scenes almost dragged one into the melee of combat, only to let go and allow the viewer to move to different sectors of the snapshot. In fact, if we adhere to this program, then the confrontation between the king and his enemy was intended to be the first segment of the event to be seen. Later, there would be the icons of citadel, marching soldiers, fleeing enemies, and the like. Cf. John Baines, "On the Status and Purposes of Ancient Egyptian Art," p. 85, and the differing viewpoint of Harold Hayden Nelson in Epigraphic Survey, *Medinet Habu I, Earlier Historical Records of Ramses III* (Chicago, 1930), pp. v–viii.

the plots, were supplied. The peeping frontal face[39] of an Asiatic enemy at Yenoam, for example, was, if not new, unexpected, as were the pictorial facets of despair and desolation in the environs of Pa-Canaan. Those small representations — a child between parents, the rapid climb into the gorse-laden hills, or even the exhausted meeting between two defeated Sinai inhabitants — point to the dramatic shift in the art. Granted that all of the latter were commonplace topoi, ones that, as with the sleepy (or inebriated) watchman, could in some way or even in a transformation be found elsewhere in Egyptian relief work then and also at an earlier time. Yet none of them may be seen in any military relief dating to the XVIIIth Dynasty. The Sinai landscape itself remains unstructured and visually a metaphorical desert.[40]

Undoubtedly, the eye of the viewer had to focus on the key object (pharaoh): size indicated that. In addition, the surrounding depictions of enemies, landscape, and fortresses played their roles in advancing the plot of the highlighted single event. I also feel that these narrative scenes provided more than eternal representations of truth (Maat) or recapitulations of an eternal combat of pharaoh versus chaos. In many ways the assumed static depiction performed the role of ekphrasis.[41] The necessity of providing a vivid description of a military event focused the artists upon one specific role of the monarch in one place and at one time. But the latter was understood by means of an aorist perspective. The viewer must traverse scene after scene in order to gather together the individual strands, "historical snapshots" if you will, into a unity. Yet any individual visual event could be understood independently of the entire pictorial story, because each scene reveals the power of its image or images whose representations were anchored in various structures.

39 The art historian Meyer Schapiro presented semiotic interpretation on the frontal view in "Frontal and Profile as Symbolic Forms," in *Words and Pictures. On the Literal and the Symbolic in the Illustration of a Text* (The Hague and Paris, 1973), chapter 4. In Egyptology, see a useful article by Christiane Favard-Meeks, "Face et profil dans l'iconographie égyptienne," *Orientalia Lovaniensia Periodica* 23 (1992): 15–36; Georges Posener, *Cinq figurines d'envoûtement* (Cairo, 1987), p. 8; and Youri Volokhine, *La frontalité dans l'iconographie de l'Egypt ancienne* (Geneva, 2000).

40 See Juan Carlos Moreno García, *Études sur l'administration, le pouvoir et l'idéologie en Égypte de l'Ancien au Moyen Empire* (Liège, 1997), p. 21; compare Alessandro Roccati, "La conception rituelle du désert chez les anciens égyptiens," in Y. Christe et al., eds., *Le désert: Image et réalité. Actes du Colloque de Cartigny 1983* (Leuven, 1983), pp. 127–29.

41 Considering the art historical and literary use of the term ekphrasis, let me refer the reader to two recently published analyses: Ruth Webb, "*Ekphrasis* Ancient and Modern: The Invention of a Genre," *Word and Image* 15 (1999): 7–18, and Graham Zanker, *Modes of Viewing in Hellenistic Poetry and Art* (Madison, 2004), chapter 1. A series of discussions may be found in Gottfried Boehm and Helmut Pfotenhauer, eds., *Beschreibungskunst – Kunstbeschreibung: Ekphrasis von der Antike bis zur Gegenwart* (Munich, 1995).

Seti I's artists had taken the initiative in providing a fresher and more historically progressive orientation to their task. The earlier narrative depictions of Hatshepsut at Deir el Bahari (Punt Reliefs) or the XVIIIth Dynasty military reliefs do not come as close as Seti I's Karnak accounts in providing a historical representation that is distinct and without widely separated stages. In this context, Kitchen has commented upon the Karnak relief blocks of Amunhotep II's campaign that they "were on granite, in vertical format. Hence, they did not come from a horizontal wall-display, but from a major doorjamb, or a doorway-thickness. In his time, there were plenty of bare walls available if wanted."[42] In other words, earlier Dynasty XVIII military depictions on walls were not as developed in narrative form as those of Seti I and his successors. This assumption, however, needs to be corrected to some degree, owing to the recent discoveries of Stephen Harvey at Abydos and W. Raymond Johnson at Karnak.[43]

The entire plan, and not merely the one associated with the Sinai, reveals a desire to provide, temporally speaking, closely connected snapshots. We do not move "to there" and "back again" as with the Egyptian vessels setting sail to Punt and then preparing to return. With Seti I, each framed event is either but a small step away from the previous or, more significantly, yet another setting designed to reflect the importance of the king.

In urban Western Asia Egyptian rulers fought in the field against fortress cities and nothing more. Distinctive in this rendering are the specific conditions of human alterity (Shasu and other barbarian enemies) and the interplay of presupposed Egyptian models. The clear separation between the Sinai and Egypt receives particular emphasis by means of the vertically-drawn canal isolating the Sinai from the homeland. This is the one case of New Kingdom pictorial representation in which "there" and "here" is made so distinct. I believe that this was accomplished by the policy of the master designer or designers. It must have been due to the radical difference between an urban civilization based on lush agriculture and the hostile desert of the immediate northeast. True, the later scenes of Ramesses III fighting against the Libyans in his eleventh regnal year may be cited to the contrary. But even there, no ecological reflection of the western desert or littoral can be discerned. To Seti I and his New Kingdom contemporaries, the Sinai was truly a land inhabited by unsettled barbarians, and these Karnak scenes prove this contention.

---

42 To a degree, this is Kitchen's argument against my position in *The Transformation of an Ancient Egyptian Narrative: P. Sallier III and the Battle of Kadesh* (Wiesbaden, 2002). See his review of that work in *JEA* 90 (2004): 45 (Reviews Supplement).

43 See most recently W. Raymond Johnson, "Tutankhamen Battle Narratives at Luxor," *KMT* 20.4 (2009): 20–33. I have discussed both important reliefs (Ahmose and Tutankhamun's) in *Icons of Power*, chapters 1 and 2.

# Bibliography

*The abbreviations employed are the standard Egyptological, Classical, and Near Eastern ones.*


Abd el-Maksoud, Mohammed. *Tell Heboua (1981–1991)* (Paris: Éditions Recherches sur les Civilisations; 1998).

——. *Enquête archéologique sur la Deuxième Période Intermédiaire et le Nouvel Empire à l'extrémité orientale du Delta* (Paris: Éditions Recherches sur les Civilisations; 1998).

——. "Tjarou, porte de l'Orient," in: Charles Bonnet and Dominique Valbelle (eds.), *Le Sinaï durant l'Antiquité et le Moyen Âge: 4,000 ans d'histoire pour un désert* (Paris: Edition Errance; 1998), 61–65.

Abd el-Maksoud, Mohammed, and Valbelle, Dominique. "Tell Héboua II. Rapport sur le décor et l'épigraphie des éléments architectoniques découverts au cours des campagnes 2008–2009 dans la zone centrale du *khétem* de Tjarou," *RdE* 62 (2011): 1–17.

Allen, James. *The Heqanakht Papyri* (New York: Metropolitan Museum of Art; 2002).

Alliot, Maurice. *Le culte d'Horus à Edfou au temps des Ptolémées I* (Cairo: Institut Français d'Archéologie Orientale; 1954).

Amer, Amin A. M. A. "Reflections on the Reign of Ramesses VI," *JEA* 71 (1985): 66–70.

Arquier, Bernard, and Guilhou, Nandine. "Cette obscure clarté qui tombe des étoiles. Les fêtes de fondation et de dédicace du temple d'Edfou," in Amanda-Alice Maravelia (ed.), *En quête de la lumière. In Quest of Light: Mélanges in Honorem Ashraf A. Sadek* (Oxford: Archaeopress; 2009), 3–18.

Assmann, Jan. *Zeit und Ewigkeit: Ein Beitrag zur Geschichte der Ewigkeit* (Heidelberg: C. Winter; 1975).

——. "Flachbildkunst des Neuen Reichs," in Claude Vandersleyen (ed.), *Das alte Ägypten* (Berlin: Propyläen Verlag; 1975), 304–25.

——. "Krieg und Frieden im alten Ägypten. Ramses II. und die Schlacht bei Kadesh," *Mannheimer Forum*, 1983/84 (1984), 175–231.

——. Steinzeit und Sternzeit. *Altägyptische Zeitkonzepte* (Munich: Wilhelm Fink; 2011).

Ayres Jr., B. Drummond, "Indicated Expert on Mideast is Found Dead in Virginia," *The New York Times* (April 30, 1983): 6.

Backer, Fabrice de. "Evolution of War Chariot Tactics in the Ancient Near East," *UF* 41 (2009): 29–45.

Baines, Alice V. "The Function of Iconography as Autobiographical Narration in the Tomb of Khnemhotep at Beni Hassan (Tomb 3)," *JNSL* 21.2 (1995): 1–24.

Baines, John. "On the Status and Purposes of Ancient Egyptian Art," *CAJ* 4 (1994): 67–94.

Bakir, Abd el-Mohsen. *The Cairo Calendar No. 86637* (Cairo: General Organisation for Government Printing Offices; 1966).

Baplu, Nele. "Lists of Months in Greek Papyri," *CdE* 83 (2008): 224–44.

Barnett, R. D. "Monkey Business," *JANES* 5 (1973): 1–5.

Baud, Michel. "The Relative Chronology of Dynasties 6 and 8," in Erik Hornung, Rolf Krauss, and David A. Warburton (eds.), *Ancient Egyptian Chronology* (Leiden and Boston: Brill; 2006), 124–43.

Baud, Michel, and Dobrev, Vassil. "De nouvelles annales de l'Ancien Empire égyptien. Un 'Pierre de Palerme' pour la VIe Dynastie," *BIFAO* 95 (1995): 23–63.

——. "Le verso des annales de la VIe Dynastie," *BIFAO* 97 (1997): 35–42.

Beal, Richard. *The Organisation of the Hittite Military* (Heidelberg: C. Winter; 1992).

Beaulieu, Paul-Alain. "The Impact of Month-lengths on the Neo-Babylonian Cultic Calendar," *ZA* 83 (1993): 66–87.

Beckerath, Jürgen von. "Das Thronbesteigungsdatum Ramses II," *ZÄS* 81 (1956): 1–3.

——. "Die 'Stele der Verbannten' im Museum des Louvre," *RdE* 20 (1968): 7–34.

——. "Ein Wunder des Amun bei der Tempelgründung in Karnak," *MDAIK* 37 (1981): 41–49.

Bedier, Shafia. "Ein Stiftungsdekret Thutmosis III.," *Bulletin of the Center of Papyrological Studies* 10 (1994): 1–23.

Bell, Lanny. "Luxor Temple and the Cult of the Royal Ka," *JNES* 44 (1985): 251–94.

——. "The New Kingdom 'Divine' Temple: The Example of Luxor," in Byron E. Schafer (ed.), *Temples of Ancient Egypt* (Ithaca; Cornell University Press; 1997), 127–84.

Bennett, Chris. "Egyptian Lunar Dates and Temple Service Months," *BiOr* 65 (2008): 525–54.

———. *Alexandria and the Moon. An Investigation into the Lunar Macedonian Calendar of Ptolemaic Egypt* (Leuven: Peeters; 2011).

Berlev, Oleg. "Bureaucrats," in Sergio Donadoni (ed.), *The Egyptians,* Robert Bianchi, trs. (Chicago: University of Chicago), 87–120.

———. "Precisely Two Sothic Periods Between Year 18 of King Sénw/Tosorthros and Year 2 of Pharaoh Antoninus Pius," in O. I. Pavlova (ed.), *Ancient Egypt: Language, Culture, Consciousness* (Moscow: Priscel's; 1999), 42–62.

———. Review of Peter der Manuelian (ed.), *Studies in Honor of William Kelly Simpson* (Boston: Dept. of Ancient Egyptian, Nubian and Near Eastern Art, Museum of Fine Arts; 1996), *BiOr* 56 (1999): 597–601.

Beylage, Peter. *Aufbau der königlichen Stelentexte vom Beginn der 18. Dynastie bis zur Amarnazeit* (Wiesbaden: Harrassowitz: 2002).

Bickerman, Elias J. *Chronology of the Ancient World*[2] (Ithaca: Cornell University Press; 1980).

Bierbrier, Morris L. *Who Was Who in Egyptology*[4] (London: Egypt Exploration Society; 2012).

Boehm, Gottfried, and Pfotenhauer, Helmut (eds.), *Beschreibungskunst – Kunstbeschreibung: Ekphrasis von der Antike bis zur Gegenwart* (Munich: Fink; 1995).

Bonneau, Danielle. "Les fêtes d'Amesysia," *CdE* 49 (1974): 366–79.

Borchardt, Ludwig. *Die Annalen und die zeitliche Festlegung des Alten Reiches der Ägyptischen Geschichte* (Berlin: Behrend & Co.; 1917).

Borghouts, J. F. *Nieuwjaar in het Oude Egypte* (Leiden: Rijksuniversiteit te Leiden; 1986).

Botti, Guiseppe. "Il carro del sogno. (Per un grato ricordo personale)," *Aegyptus* 31 (1951): 192–98.

Boulanger, Jean-Paul, and Rensio, Geneviève. *Naissance de l'écriture: cunéiformes et hiéroglyphes: [exposition], Galeries nationales du Grand Palais, 7 mai–9 août 1982* (Paris: Ministère de la culture, Editions de la réunion des musées nationaux; 1982).

Bourriau, Janine. "Three Monuments from Memphis in the Fitzwilliam Museum," *JEA* 62 (1982): 51–59.

Bowen, Alan C., and Goldstein, Bernard R. "Meton of Athens and Astronomy in the Late Fifth Century B.C.," in Erle Leichty, Maria deJ. Ellis, and Pamela

Gerardi (eds.), *A Scientific Humanist: Studies in Honor of Abraham Sachs* (Philadelphia: University Museum; 1988), 39–81.

———. "On Early Hellenistic Astronomy," *Centaurus* 32 (1989): 272–93.

———. "The Introduction of Dated Observations and Precise Measurement in Greek Astronomy," *Archive for History of Exact Sciences* 43 (1991): 93–132.

Brack-Bernsen, Lis. "The 360-Day Year in Mesopotamia," in John M. Steele (ed.), *Calendars and Years: Astronomy and Time in the Ancient Near East* (Oxford: Oxbow; 2007), 83–100.

Brand, Peter. *The Monuments of Seti I: Epigraphic, Historical, and Art Historical Analysis* (Leiden and Boston: Brill; 2000).

———. "The Date of the War Scenes on the South Wall of the Great Hypostyle Hall and the West Wall of the Cour de la Cachette at Karnak and the History of the Late Nineteenth Dynasty," in Mark Collier and Steven Snape (eds.), *Ramesside Studies in Honour of K.A. Kitchen* (Bolton: Rutherford Press; 2011), 51–84.

Breasted, James Henry. "The Battle of Kadesh: A Study in the Earliest Known Military Strategy," in *Decennial Publications of the University of Chicago* V (Chicago: University of Chicago Press, 1903), 81–126.

———. *The Oriental Institute*[2] (Chicago: Oriental Institute; 1933).

Broadhurst, Clive. "An Artistic Interpretation of Sety I's War Reliefs," *JEA* 75 (1989): 229–34.

Brugsch, Heinrich. "Bau und Maaßse des Temples von Edfu," *ZÄS* 10 (1872): 1–16.

Bryan, Betsy. "Property and God's Wives of Amun," Presentation at a conference called "Women and Property," organized by D. Lyons and R. Westbrook at Harvard University, 2005.

Bryce, Trevor. *Hittite Warrior* (Osprey: Oxford and New York; 2007).

Burne, Alfred H. "Some Notes on the Battle of Kadesh: Being a Military Commentary on Professor J.H. Breasted's Book, 'The Battle of Kadesh'" (University of Chicago Press, 1903), *JEA* 7 (1921): 191–95.

Cameron, George. "Darius and Xerxes in Babylonia," *AJSL* 58 (1941): 314–25.

———. "Darius, Egypt and 'The Lands Beyond the Sea,'" *JNES* 2 (1943): 307–13.

Caminos, Ricardo. *Semna-Kumma I: The Temple of Semna* (London: Egypt Exploration Society; 1998).

Caritoux, Laurence. *Le nom des attelages royaux au Nouvel Empire* (Montpellier, MA Thesis: University of Montpellier; 1996).

———. "Les chevaux de Pharaoh," *Égypte. Afrique et Orient* 11 (1998): 21–26.

Carlotti, Jean-François. *L'Akh-menou de Thoutmosis III à Karnak: étude architecturale* (Paris: Recherche sur les Civilisations; 2001).

Carver, Charles Hinson. *The Egyptian Chariot* (Providence, PhD Dissertation: Brown University; 1955).

Cauville, Sylvie, and Devauchelle, Didier. "Le temple d'Edfou: étapes de construction. Nouvelles données historiques," *RdE* 35 (1984): 31–55.

Cavillier, Giacomo. *La battaglia di Qadesh: Ramesse II alla conquista dell'Asia, fra mito, storia e strategia* (Turin: Tirrenia Stampatori; 2006).

Cenival, J. L. de. "Un nouveau fragment de la pierre de Palerme," *BSFE* 44 (1965): 13–17.

Černý, Jaroslav. "The Origin of the Name of the Month Tybi," *ASAE* 43 (1943): 173–81.

———. "Philological and Etymological Notes," *ASAE* 51 (1951): 441–46.

———. "Some Coptic Etymologies III," *BIFAO* 57 (1958): 203–13.

Clagett, Marshall. *Ancient Egyptian Science: Calendars, Clocks, and Astronomy* (Philadelphia: American Philosophical Society; 1995).

Cohen, Beth (ed.). *Not the Classical Ideal: Athens and the Construction of the Other in Greek Art* (Leiden: Brill; 2000).

Cotterell, Brian, Dickson, F. P., and Kamminga, Johann. "Ancient Egyptian Water-clocks: A Reappraisal," *Journal of Archaeological Science* 13 (1986): 31–50.

Daborn, Joseph. *Military Use of the Horse in the New Kingdom* (Swansea, Undergraduate Dissertation: University of Wales, Swansea; 2005).

Darnell, John Coleman. "Supposed Depictions of Hittites in the Amarna Period," *SAK* 18 (1991): 113–40.

———. *Theban Desert Road Survey in the Egyptian Western Desert. Volume 1: Gebel Tjauti Rock Inscriptions 1–45 and Wadi el-Hôl Rock Inscriptions 1–45* (Chicago: Oriental Institute; 2002).

Dawson, Warren R. "An Oracle Papyrus B.M. 10335," *JEA* 11 (1925): 247–48.

Davies, Nina de Garis, and Gardiner, Alan H. *The Tomb of Amenemhet (No. 82)* (London: Egypt Exploration Fund; 1915).

Delbrück, Hans. *Numbers in History* (London: University of London; 1913).

———. *Warfare in Antiquity,* Walter J. Renfroe, Jr., trs. (Lincoln and London: University of Nebraska Press; 1990).

Degrève, Agnès. "La campagne asiatique de l'an 1 de Séthy Ier représentée sur le mur extérieur nord de la salle hypostyle du temple d'Amon à Karnak," *RdE* 57 (2006): 47–76.

Department of the Army, Headquarters. *Foot Marches* (Field Manuel No. 21-18) (Washington DC: Department of the Army, USA; 1990).

Depuydt, Leo. "On the Consistency of the Wandering Year as Backbone of Egyptian Chronology," *JARCE* 32 (1995): 43–58.

———. *Civil Calendar and Lunar Calendar in Ancient Egypt* (Leuven: Peeters; 1997).

———. "The Demotic Mathematical Astronomical Papyrus Carlsberg 9 Reinterpreted," in W. Clarysse, A. Schoors, and H. Willems (eds.), *Egyptian Religion: The Last Thousand Years. Studies Dedicated to the Memory of Jan Quaegebeur* II (Leuven: Peeters; 1998), 1277–97.

———. "The Two Problems of the Month Names," *RdE* 50 (1999): 107–33.

———. "What is Certain about the Origin of the Egyptian Civil Calendar?" in Hedvig Győry (ed.), *Mélanges offerts à Edith Varga: "Le lotus qui sort de terre"* (Budapest: Musée Hongrois des Beaux-Arts, 2001), 81–94.

———. "Esna's Triple New Year," *JARCE* 40 (2003): 55–67.

———. "Calendars and Years in Ancient Egypt," in John M. Steele (ed.), *Calendars and Years: Astronomy and Time in the Ancient Near East* (Oxford: Oxbow; 2007), 35–81.

Derchain, Philippe. "La clepsydre de la dernière nuit," *GM* 210 (2006): 19–21.

Devauchelle, Didier. "Écrire le nom des jours épagomènes et du premier jour de l'an [O. Dém Delm 4-1]," *SEP* 2 (2005): 75–81.

Doggett, LeRoy E., and Schaefer, Bradley. "Lunar Crescent Visibility," *Icarus* 107 (1994): 388–403.

Eaton, Katherine. "Monthly Lunar Festivals in the Mortuary Realm: Historical Patterns and Symbolic Motifs," *JNES* 70 (2011): 229–45.

Eaton-Krauss, Marianne. Review of William Stevenson Smith, *The Art and Architecture of Ancient Egypt* (New Haven; Yale University Press; 1998), *OLZ* 95 (2000): 396–400.

Edwards, T. E. S. *From the Pyramids to Tutankhamun: Memoirs of an Egyptologist* (Oxford: Oxbow; 2000).

El-Razik, Mahmud Abd. "The Dedicatory Inscription and Building Texts of Ramesses II in Luxor Temple," *JEA* 61 (1975): 125–36.

El-Sabban, Sherif. *Temple Festival Calendars of Ancient Egypt* (Liverpool: Liverpool University Press; 2000).

Epigraphic Survey. *Medinet Habu I, Earlier Historical Records of Ramses III* (Chicago: University of Chicago Oriental Institute; 1930).

———. *The Battle Reliefs of King Sety I, Reliefs and Inscriptions at Karnak IV* (Chicago: University of Chicago oriental Institute; 1985).

Eshmawy, Aiman. "Names of Horses in Ancient Egypt," in Jean-Claude Goyon and Christine Cardin (eds.), *Proceedings of the Ninth International Congress of Egyptologists* I (Leuven, Paris, and Dudley, MA: Peeters; 2007), 665–76.

Essche, Éric van. "Les félins à la guerre," in L. Delvaux and E. Warmenbol (eds.), *Les divins chats d'Égypte. Un air subtil. Un dangereux parfum* (Leuven: Peeters; 1991), 31–48.

Essche-Merchez, Éric van. "Quelques réflexions sur l'espace et le récit à Médinet Habou," *Annales d'Histoire de l'Art et Archéologie* 11 (1989): 7–24.

———. "Scène, paroi, récit. Des cases de bande dessinée aux reliefs ramessides: pour une complémentarité méthodologique," *Art & Fact* 13 (1994): 40–57.

———. "Pour une lecture 'stratigraphique' des parois du temple de Ramsès III à Médinet Habou," *RdE* 45 (1994): 87-116.

Faulkner, Raymond O. "The Battle of Kadesh," *MDAIK* 16 (1958): 93–111.

Favard-Meeks, Christiane. *Words and Pictures. On the Literal and the Symbolic in the Illustration of a Text* (The Hague and Paris: Mouton; 1973).

Fecht, Gerhard. "Das 'Poem' über die Qadeš-Schlacht," *SAK* 11 (1984): 281–333.

Feldman, Marian H., and Sauvage, Caroline. "Objects of Prestige? Chariots in the Late Bronze Age Eastern Mediterranean and Near East," *ÄuL* 20 (2010): 67–181.

Fermor, John. "Perceived Night Length Ratios in Ancient Egypt," *Vistas in Astronomy* 36 (1993): 363–73.

Figueras, Pau. *From Gaza to Pelusium: Materials for the Historical Geography of North Sinai and Southwestern Palestine (332 BCE–640 CE)* (Beer Sheva: Ben-Gurion University of the Negev Press; 2000).

Fischer, Henry George. *The Orientation of Hieroglyphs: Part I. Reversals* (New York: Metropolitan Museum of Art; 1977).

———. *L'écriture et l'art de l'Égypte ancienne: quatre leçons sur la paléographie et l'épigraphie pharaoniques* (Paris: Presses universitaires de France; 1986).

———. *Varia Nova (Egyptian Studies III)* (New York: Metropolitan Museum of Art; 1996).

———. "Notes on Some Texts of the Old Kingdom and Later," in Peter der Manuelian (ed.), *Studies in Honor of William Kelly Simpson* I (Boston: Dept. of Ancient Egyptian, Nubian and Near Eastern Art, Museum of Fine Arts; 1996), 269–74.

Fotheringham, J. K. "The Probable Error of a Water Clock," *The Classical Review* 29 (1915): 236–38.

Francia, Pier Roberto del. "Il carro di Firenzi," in Maria Christina Guidotti (ed.), *Il carro e le armi del Museo egizio di Firenze* (Florence: Giunti; 2002), 16–37.

Frandsen, Paul John. *An Outline of the Late Egyptian Verbal System* (Copenhagen: Akademisk Forlag; 1974).

Franke, Detlef. "Zur Chronologie des Mittleren Reiches, Teil 1: Die 12. Dynastie," *Or* 57 (1988): 113–38.

———. "Zur Chronologie des Mittleren Reiches, Teil II: Die sogenannte 'Zweite Zwischenzeit Altägyptens,'" *Or* 57 (1988): 245–74.

Frankfort, Henri. *The Cenotaph of Seti I at Abydos* II (London: Egypt Exploration Society; 1933).

Fukaya, Masashi, "Oracular Sessions and the Installations of Priests and Officials at the Opet Festival," *Orient* 47 (2012): 191–211.

———. *Socio-Religious Functions of Three Theban Festivals in the New Kingdom: The Festivals of Opet, the Valley, and the New Year* (Oxford, PhD Dissertation: Oxford University; 2014).

Gaballa, G. A. *Narrative in Egyptian Art* (Mainz am Rhein: Philipp von Zabern; 1976).

Gardiner, Alan H. "Mesore as the First Month of the Egyptian Year," *ZÄS* 43 (1906): 136–44.

———. Review of J. G. Frazer, *The Golden Bough: Adonis, Attis, Osiris; Studies in the History of Oriental Religion* (London: Macmillan and Co.; 1914), *JEA* 2 (1915): 121–26.

———. "Regnal Years and Civil Calendar in Pharaonic Egypt," *JEA* 31 (1945): 11–28.

———. "Tuthmosis III Returns Thanks to Amun," *JEA* 38 (1952): 6–23.

———. "The Problem of the Month-Names," *RdE* 10 (1955): 9–31.

———. *Egyptian Grammar*[3] (Oxford; Griffith Institute; 1957).

———. *The Kadesh Inscriptions of Ramesses II* (Oxford: Griffith Institute; 1960).

Gauthier, Henri. "Quatre nouveaux fragments de la pierre de Palerme," *Le Musée Égyptien* 3 (1915): 29–53.

———. *Les fêtes du dieu Min* (Cairo: Institut Français d'Archéologie Orientale; 1931).

Gautschy, Rita. "Lunar and Sothic Data from the Archive of el-Lahun Revisited: Chronology of the Middle Kingdom," in Maarten Horn et al. (eds.), *Current Research in Egyptology 2010: Proceedings of the Eleventh Annual Symposium* (Oxford and Oakville: Oxbow; 2011), 53–56.

———. "Monddaten aus dem Archiv von Illahun: Chronologie des Mittleren Reiches," *ZÄS* 138 (2011): 1–19.

Gilroy, Thomas G. "Outlandish Outlanders: Foreigners and Caricature in Egyptian Art," *GM* 191 (2002): 35–52.

Gitton, Michel. *L'épose du dieu Ahmès Néfertary: documents sur sa vie et son culte posthume* (Paris: Les Belles Lettres; 1981).

———. Les divines épouses de la 18e Dynastie (Paris: Les Belles Lettres; 1984).

Gnirs, Andrea Maria. *Militär und Gesellschaft: Ein Beitrag zur Sozialgeschichte des Neuen Reiches* (Heidelberg: Heidelberger Orientverlag; 1996).

Goedicke, Hans. "A Cult Inventory of the Eighth Dynasty from Coptos (Cairo JE 43290)," *MDAIK* 50 (1994): 71–84.

Gohary, Jocelyn. *Akhenaton's Sed-Festival at Karnak* (London and New York: Kegan Paul International; 1992).

Gozzoli, Roberto B. *The Writing of History in Ancient Egypt during the First Millennium BC (ca. 1070–180 BC): Trends and Perspectives* (London: Golden House Publications; 2009).

Grandet, Pierre. *Le Papyrus Harris I (BM 9999)* (Cairo: Institut Français d'Archéologie Orientale; 1994).

Griffith, F. Ll. *Two Hieroglyphic Papyri from Tanis* (London: Trübner & Co; 1889).

Grimal, Nicholas. "La danse des peuples aux marches du royaume," *CRAIBL* (2001): 1159–83.

Grimm, Alfred. Die altägyptischen Festkalender in den Tempeln der griechisch-römischen Epoche (Wiesbaden: Harrassowitz; 1994).

———. "Zur kalendarischen Fixierung des *jhhy*-(Freuden-)Festes nach dem Festkalender des Königs Amenophis I. aus Karnak," *GM* 143 (1994): 73–76.

Guidotti Maria Christiana (ed.). *Il carro e le armi del Museo egizio di Firenze* (Florence: Giunti; 2002).

Hagen, Frederik, and Koefoed, Henriette. "Private Feasts at Deir el-Medina: Aspects of Eating and Drinking in an Ancient Egyptian Village," *Archaeological Review from Cambridge* 20 (2005): 6–31.

Hamblin, William. *The Fatimid Army during the Early Crusades* (Ann Arbor, PhD Dissertation: University of Michigan; 1985).

Hannah, Robert. *Greek and Roman Calendars. Constructions of Time in the Classical World* (London: Duckworth; 2005).

Hannig, Rainer. *Großes Handwörterbuch Ägyptisch–Deutsch (2800–950 v. Chr.): Die Sprache der Pharaonen* (Mainz: Philipp von Zabern, 1995).

Harari, Ibrahim. "Nature de la stèle de donation de fonction du roi Ahmôsis à la reine Ahmès-Nefertari," *ASAE* 56 (1959): 139–201.

Hare, Tom. *Remembering Osiris: Number, Gender and the Word in Ancient Egyptian Representational Systems* (Stanford: Stanford University Press; 1999).

Haring, Ben J. J. *Divine Households: Administrative and Economic Aspects of the New Kingdom Royal Memorial Temples in Western Thebes* (Leiden: Nederlands Instituut voor het Nabije Oosten; 1997).

Harrison, Graeme. "The Position of the Day *grḥ nẖnw m sš.f* — 'The Night the Child is in his Nest' within the Epagomenal Period," *GM* 143 (1994): 77–79.

Hasel, Michael G. "The Battle of Kadesh: Identifying New Kingdom Polities, Places, and Peoples in Canaan and Israel," in S. Bar, D. Kahn, and J. J. Shirley (eds.), *Egypt, Canaan and Israel: History, Imperialism, Ideology and Literature* (Leiden and Boston: Brill; 2011), 65–85.

Hayes, William C. "Inscriptions from the Palace of Amenhotep III," *JNES* 10 (1951): 35–56, 82–111, 156–83, and 231–42.

Heagren, Brett H. *The Art of War in Pharaonic Egypt: An Analysis of the Tactical, Logistic, and Operational Capabilities of the Egyptian Army (Dynasties XVII–XX)* (Auckland, PhD Dissertation: University of Auckland; 2010).

Heinz, Susanne Constanze. *Die Feldzugsdarstellungen des Neuen Reiches. Eine Bildanalyse* (Vienna: Akademie der Wissenschaften; 2001).

Helck, Wolfgang. "Ramessidische Inschriften aus Karnak," *ZÄS* 82 (1956): 98–109.

——. *Materialien zur Wirtschaftsgeschichte des Neuen Reiches* III. Akademie der Wissenschaften und der Literatur, Abhandlungen der Geistes- und Sozialwissenschaftlichen Klasse 1962.2 (Wiesbaden: F. Steiner; 1962).

——. "Feiertage und Arbeitstage in der Ramessidenzeit," *JESHO* 7 (1964): 136–66.

——. "Überlegungen zur Geschichte der 18. Dynastie," *OrAnt* 8 (1969): 281–327.

——. "Zur Opferliste Amenophis' IV. (*JEA* 57, 70ff.)," *JEA* 59 (1973): 95–99.

——. *Historisch-biographische Texte der 2. Zwischenzeit und neue Texte der 18. Dynastie* (Wiesbaden: Harrassowitz; 1975).

——. "Die 'Weihinschrift' aus dem Taltempel des Sonnenheiligtums des Königs Neuserre bei Abu Gurob," *SAK* 5 (1977): 47–77.

Henne, Willibald. "Bemerkungen zur altägyptischen Streiflicht-Uhr," *ZÄS* 118 (1991): 21–31.

Hikade, Thomas. "Crossing the Frontier into the Desert: Egyptian Expeditions to the Sinai Peninsula," *AWE* 6 (2007): 1–22.

Hintze, Fritz, et al. *Musawwarat es Sufra: Der Löwentempel. Textband* (Berlin: Akademie Verlag; 1993).

Hoffmeier, James K., and Abd el-Maksoud, Mohammed. "A New Military Site on the 'Ways of Horus:' Tell el-Borg 1999–2001: A Preliminary Report," *JEA* 89 (2003): 169–97.

———. *Ancient Israel in the Sinai: The Evidence for the Authenticity of the Wilderness Tradition* (New York: Oxford University Press; 2005).
———. "The Gate of the Ramesside Fort at Tell-Bourg, North Sinai," in Mark Collier and Steven Snape (eds.), *Ramesside Studies in Honour of K. A. Kitchen* (Bolton: Rutherford Press; 2011), 207–18.
———. *Tell el-Borg I: Excavations in North Sinai* (Winona Lake: Eisenbrauns; 2014).

Hoffmeier, James, and van Dijk, Jacobus. "New Light on the Amarna Period from North Sinai," *JEA* 96 (2010): 191–205.

Hoffmeier, James, and Moshier, Stephen. "'A Highway out of Egypt:' The Main Road from Egypt to Canaan," in: Frank Förster and Heiko Riemer (eds.), *Desert Road Archaeology in Ancient Egypt and Beyond* (Cologne; Heinrich-Barth-Institut; 2013), 485–510.

Hornung, Erik. *Nacht und Finsternis im Weltbild der Alten Ägypter* (Tübingen, PhD Dissertation: Eberhard-Karls-Universität; 1956).
———. "Pharao Ludens," *Eranos Jahrbuch* 51 (1982): 479–516.
———. "The New Kingdom," in Erik Hornung, Rolf Krauss, and David Warburton (eds.), *Ancient Egyptian Chronology* (Leiden and Boston: Brill; 2006), 197–217.

Hornung, Erik, and Staehlin, Elisabeth. *Neue Studien zum Sedfest* (Basel: Schwabe; 2006).

Hubaux, Jean, and Leroy, Maxime. *Le mythe du phénix dans les littératures grecque et latine* (Paris: Faculté de Philosophie et Lettres, Liège; 1939).

Huber, Peter J. "Modeling the Length of Day and Extrapolating the Rotation of the Earth," in Fabrizio Bònili, Salvo de Meis, and Antonio Panaino (eds.), *Astronomical Amusements: Papers in Honor of Jean Meeus* (Milan: Mimesis; 2000), 91–104.

Hudson, Grover. "Ethiopian Month-names," in Gábor Takács (ed.), *Semito-Hamitic Festschrift for A. B. Dolgopolsky and H. Jungraithmayr* (Berlin: Dietrich Reimer Verlag; 2008), 150–54.

Ingham, M. F. "The Length of the Sothic Cycle," *JEA* 55 (1969): 36–40.

James, Thomas Garnet Henry. "The Date of the Month *rkḥ wr*," *JEA* 41 (1955): 123.
———. *The Ḥekanakhte Papers and Other Early Middle Kingdom Documents* (New York: Metropolitan Museum of Art; 1962), 3–4.

Janssen, Jac. J. "Rations with Riddles," *GM* 124 (1991): 91–97.
———. "Rations with Riddles II," *GM* 128 (1992): 81–86.

Jansen-Winkeln, Karl. *Inschriften der Spätzeit. Teil I: Die 21. Dynastie* (Wiesbaden: Harrassowitz, 2007.

——. *Inschriften der Spätzeit. Teil II: Die 22.–24. Dynastie* (Wiesbaden: Harrassowitz; 2007).

Jasnow, Richard. "On Sothis and the So-Called Clepsydra (?) in the Book of Thoth (B02, 3/16-4/11)," in Hermann Knuf, Christian Leitz, and Daniel von Recklinghausen (eds.), *Honi soit qui mal y pense: Studien zum pharaonischen, griechisch-römischen und spätantiken Ägypten zu Ehren von Heinz-Josef Thissen* (Leuven, Paris, and Walpole MA: Peeters; 2010), 237–43.

Jauhiainen, Heidi. *"Do not Celebrate your Feast without your Neighbours": A Study of References to Feasts and Festivals in Non-Literary Documents from Ramesside Period Deir el-Medina* (Helsinki, PhD Dissertation, University of Helsinki: University Press; 2009).

Johnson, W. Raymond. "Tutankhamen Battle Narratives at Luxor," *KMT* 20.4 (2009): 20–33.

Jones, Alexander. "On the Reconstructed Macedonian and Egyptian Lunar Calendar," *ZPE* 119 (1997): 157–66.

Jong, Teije de. "The Heliacal Rising of Sothis," in Erik Hornung, Rolf Krauss, and David Warburton (eds.), *Ancient Egyptian Chronology* (Leiden and Boston: Brill; 2006), 32–38.

Kemp, Barry. *Ancient Egypt: Anatomy of a Civilization* (London and New York: Routledge; 1989).

Kenning, Jürgen. *Der Feldzug nach Kadesch: Das Ägypten des Neuen Reiches auf der Suche nach seiner Strategie* (Hildesheim; Olms-Verlag; 2014).

Kitchen, Kenneth A., and Gaballa, G. A. "The Festival of Sokar," *Or* 38 (1969): 1–76.

——. "Memoranda on Craftsmen at the Ramesseum," in Elizabeth Goring, Nicholas Reeves, and John Ruffle (eds.), *Chief of Seers: Egyptian Studies in Honour of Cyril Aldred* (London and New York: Kegan Paul; 1997), 175–77.

——. *Ramesside Inscriptions: Historical and Biographical. Notes and Comments* II (Blackwell: Oxford and Malden; 1997).

——. Review of Anthony Spalinger, *The Transformation of an Ancient Egyptian Narrative: P. Sallier III and the Battle of Kadesh* (Wiesbaden; Harrassowitz; 2002), *JEA* 90 (2004): 45 (Reviews Supplement).

Kormyscheva, Eleonora. "Festkalender im Kawa-Tempel," in Rolf Gundlach and Matthias Rochholz (eds.), *4. Ägyptologische Tempeltagung: Feste im Tempel* (Wiesbaden: Harrassowitz; 1998), 77–89.

Kóthay, Katalin Anna. "Phyles of Stone-Workers in the Phylae System of the Middle Kingdom," *ZÄS* 134 (2007): 138–50.

Krauss, Rolf. *Probleme des altägyptischen Kalenders und der Chronologie des Mittleren und Neuen Reiches in Ägypten* (Berlin, PhD Dissertation: Freie Universität; 1981).

———. *Sothis- und Monddaten: Studien zur astronomischen und technischen Chronologie Altägyptens* (Hildesheim: Gerstenberg Verlag; 1985).

———. "Was wäre, wenn der altägyptische Kalendertag mit Sonnenaufgang begonnen hätte?," *BSEG* 17 (1993): 63–71.

———. "Fällt im Illahun-Archiv der 15. Mondmonatstag auf den 16. Mondmonatstag?," *GM* 138 (1994): 81–92.

———. Review of Ulrich Luft, *Die chronologische Fixierung des ägyptischen Mittleren Reiches nach dem Tempelarchiv von Illahun* (Vienna: Akademie der Wissenschaften; 1992), *OLZ* 89 (1994): 6–18.

———. "Wenn und Aber: Das Wag-Fest und die Chronologie des Alten Reiches," *GM* 162 (1998): 43–63.

———. "Lunar Dates," in Erik Hornung, Rolf Krauss, and David Warburton (eds.), *Ancient Egyptian Chronology* (Leiden and Boston: Brill; 2006), 395–431.

———. "Egyptian Sirius/Sothic Dates, and the Question of the Sothis-Based Lunar Calendar," in Erik Hornung, Rolf Krauss, and David A. Warburton (eds.), *Ancient Egyptian Chronology* (Leiden and Boston: Brill; 2006), 439–57.

———. "Die Mahler-Borchardtsche These über die Apis-Inthronisationen bei Vollmond," *Acta Praehistorica et Archaeologica* 39 (2007): 339–48.

———. "Egyptian Calendars," *KBN* 3 (2008–9): 105–13.

———. "Astronomical Chronology," in Juan Antonio Belmonte and Mosalam Shaltout (eds.), *In Search of Cosmic Order: Selected Essays on Egyptian Archaeoastronomy* (Cairo: Supreme Council of Antiquities; 2009), 135–54.

———. "Babylonian Crescent Observations and Ptolemaic Lunar Dates," *PalArch* 9 (2012): 1–95.

Kruchten, Jean-Marie. *Le grand texte oraculaire de Djéhoutymose intendant du domain d'Amon sous le pontificat de Pinedjem* II (Brussels: Fondation Égyptologique Reine Élisabeth; 1986).

———. *Les annales des prêtres de Karnak (XXIe–XXIIIe dynasties) et autres textes contemporains relatifs à l'initiation des prêtres d'Amon* (Leuven: Department Orientalistiek; 1989).

———. "L'année où la fête d'Opet n'eut pas lieu en Paophi," *JEA* 77 (1991): 181–83.

Lacau, Pierre, and Chevrier, Henri. *Une chapelle d'Hatshepsout à Karnak* (Cairo: Institut Français d'Archéologie Orientale; 1977).

Lefèvre, D. "Egyptian Queens: A Selection of Sources," in Christiane Ziegler (ed.), *Queens of Egypt: From Hetepheres to Cleopatra* (Monaco and Paris: Grimaldi Forum and Somogy Éditions d'Art; 2008), 394–410.

Legrain, Georges. *Statues et statuettes de rois et de particuliers* III (Cairo: Institut Français d'Archéologie Orientale; 1914).

Leitz, Christian. *Studien zur ägyptischen Astronomie* (Wiesbaden: Harrassowitz; 1989).

———. *Tagewählerei: Das Buch ḥ3t nḥḥ pḥ.wy ḏt und verwandte Texte.* Tafelband (Wiesbaden: Harrassowitz; 1994).

———. Review of Leo Depuydt, *Civil Calendar and Lunar Calendar in Ancient Egypt* (Leuven: Peeters; 1997), *BiOr* 57 (2000): 75–81.

Lichtheim, Miriam. *Ancient Egyptian Literature* (Berkeley, Los Angeles, and London: University of California Press; 1976).

Lieven, Alexandra von. *Grundriss des Laufes der Sterne: Das sogenannte Nutbuch* (Copenhagen: Museum Tusculanum Press; 2007).

———. "Bemerkungen zum Dekorationsprogramm des Osireion in Abydos," in Ben J. J. Haring and Andrea Klug (eds.), *6. Ägyptologische Tempeltagung: Funktion und Gebrauch altägyptischer Tempelräume, Leiden 4.–7. September 2002* (Wiesbaden: Harrassowitz; 2007), 167–86.

———. "The Movement of Time. News from the 'Clockmaker' Amenemhet," in Renata Landgráfová and Jana Mynářová (eds.), *Rich and Great. Studies in Honor of Anthony J. Spalinger on the Occasion of his 70th Feast of Thoth* (Prague: Charles University in Prague, Faculty of Arts; 2016), 187–206.

Lippert, Sandra L. "Au clair de la lune — The Organisation of Cultic Service by Moon Calendar in Soknopaiou Nesos," in Ghislaine Widmer and Didier Devauchelle (eds.), *Actes du IXe Congrès International des Études Démotiques; Paris, 31 août–3 septembre 2005* (Cairo: Institut Français d'Archéologie Orientale; 2009), 183–94.

Littauer, Mary A., and Crouwel, Johan. *Chariots and Related Equipment from the Tomb of Tutankhamun* (Oxford: Griffith Institute; 1985).

Liverani, Mario. "The Ideology of the Assyrian Empire," in Mogens Trolle Larsen (ed.), *Power and Propaganda: A Symposium on Ancient Empires* (Copenhagen: Akademisk Forlag; 1979), 297–317.

———. "Stereotipi della lingua 'altra' nell'Asia anteriore antica," *VO* 3 (1980): 15–31.

———. *Prestige and Interest: International Relations in the Near East ca. 1600–1100 B.C.* (Padova: Sargon; 1990).

Long, Charlotte R. *The Twelve Gods of Greece and Rome* (Leiden, New York, Copenhagen, and Cologne: Brill; 1987).

Loprieno, Antonio. "Zahlwort," in Wolfgang Helck and Eberhard Otto (eds.), *Lexikon der Ägyptologie* (Wiesbaden: Harrassowitz; 1986), 1306–19.

———. *Topos und Mimesis: Zum Ausländer in der ägyptischen Literatur* (Wiesbaden: Harrassowitz; 1988);

———. *La pensée et l'écriture: pour une analyse sémiotique de la culture égyptienne. Quatre séminaires à l'École Pratique des Hautes Études, Section des sciences religieuses, 15–27 mai 2000* (Paris: Cybèle; 2001).

Luft, Ulrich. "Illahunstudien IV: Zur chronologischen Verwertbarkeit des Sothisdatums," *SAK* 16 (1989): 217–33.

———. "Das *s*ꜣ*ḏ*-Fest im Alten Reich," in Daniela Mendel und Ulrike Claudi (eds.), *Ägypten im afro-orientalischen Kontext: Aufsätze zur Archäologie, Geschichte und Sprache eines unbegrenzten Raumes: Gedenkschrift Peter Behrens* (Cologne: Institut für Afrikanistik; 1991), 237–44.

———. *Die chronologische Fixierung des ägyptischen Mittleren Reiches nach dem Tempelarchiv von Illahun* (Vienna: Akademie der Wissenschaften; 1992).

———. "The Date of the *W*ꜣ*gy* Feast: Considerations on the Chronology of the Old Kingdom," in Anthony Spalinger (ed.), *Revolutions in Time: Studies in Ancient Egyptian Calendrics* (San Antonio: Van Siclen Press; 1994), 39–44.

———. "'Weil ... nicht sein kann, was nicht sein darf:' Replik einer Kritik," *GM* 141 (1994): 109–11.

———. "Rund um den Beginn des ägyptisches Tages," in Ernst Czerny et al. (eds.), *Timelines: Studies in Honour of Manfred Bietak* I (Leuven, Paris, and Dudley: Peeters; 2006), 207–15.

Lurson, Benoît. "La légitimation du pouvoir royal par l'observance des rites osiriens: Analyse d'une séquence de scènes de la grande salle hypostyle de Karnak," *Acta Orientalia Belgica* 15 (2001): 303–32.

———. "La conception du décor d'un temple au début du règne de Ramsès II: Analyse du deuxième registre de la moitié sud du mur ouest de la grande salle hypostyle de Karnak," *JEA* 91 (2005): 107–24.

Maisler (= Mazar), Bernard. "Meḥqarîm topograpiyyîm: 5. L[e]bô' Ḥ[a]mat = Lebo Hamath and the Northern Boundary of Canaan," *Bulletin of the Jewish Israel Palestine Exploration Society* 12 (1946): 91–100.

Manuelian, Peter der. *Studies in the Reign of Amenophis II* (Hildesheim: Gerstenberg Verlag; 1987).

Marcus, Michelle I. "Geography as an Organizing Principle in the Imperial Art of Shalmaneser III," *Iraq* 49 (1987): 77–90.

———. "Geography as Visual Ideology: Landscape, Knowledge, and Power in Neo-Assyrian Art," in Mario Liverani (ed.), *Neo-Assyrian Geography* (Rome: Sargon; 1995), 193–202.

Marcus, Stephen. *Engels, Manchester, and the Working Class* (New York: Random House; 1974).

Markey, T. L. "Morning, Evening, and the Twilight Between," in Susan Nacev Skomal and Edgar C. Polomé (eds.), *Proto-Indo-European. The Archaeology of a Linguistic Problem: Studies in Honor of Marija Gimbutas* (Washington D.C.: Institute for the Study of Man; 1987), 299–321.

Matthiessen, Francis Otto. *American Renaissance: Art and Expression in the Age of Emerson and Whitman* (Oxford: Oxford University Press; 1940).

Mazar, Bernard. "Lebo-*ḥamath* and the Northern Border of Canaan," in Shmuel Ahituv and Baruch A. Levine (eds.), *The Early Biblical Period: Historical Essays* (Jerusalem: Israel Exploration Society; 1986), 189–202.

Meeks, Dimitri. *Année Lexicographique: Égypte ancienne* III (Paris: Impr. de la Margeride; 1982).

Meeus, Jean. *Astronomical Algorithms* [2] (Richmond VA: Willmann-Bell; 1999).

Mengoli, Pierangelo. "La clessidra egizia del Museo Barracco," *VO* 6 (1986): 193–209.

——. "Some Considerations of Egyptian Star-Clocks," *Archiv der Geschichte der Naturwissenschaften* 22–24 (1988): 1127–50.

——. "La clessidra di Karnak: L'orologio ad acqua di Amenophis III," *OA* 28 (1989): 227–71.

——. *Astronomica Egizia* (Budrio: private printing; 2006).

Meyer-Dietrich, Erika. "Die Opetprozession – mehr als nur eine rituelle Verbindung von Karnak mit Luxor: Die Verwendung von Schall zur Erzeugung eines symbolischen Raumes bei der Opetprozession," in I. M. Dolinska and H. Beinlich (eds.) *Interconnections between Temples* (Wiesbaden: Harrassowitz), 123–36.

Mikhail, Louise. "The Festival of Sokar: An Episode of the Osirian Khoiak Festival," *GM* 82 (1984): 25–44.

Moers, Gerald. *Fingierte Welten in der ägyptischen Literatur des 2. Jahrtausends v. Chr.: Grenzüberschreitung, Reisemotiv und Fiktionalität* (Leiden and Boston: Brill; 2001).

Moreno García, Juan Carlos. *Études sur l'administration, le pouvoir et l'idéologie en Égypte de l'Ancien au Moyen Empire* (Liège: C.I.P.L.; 1997).

Morenz, Ludwig D. "Der Mond. Über Königsnamen, Visuelle Poesie, Metaphorik und Zeitgeist in der Mitte des 2. Jt. v. Chr.," *BSEG* 24 (2002–3): 121–30.

Morris, Ellen Fowles. *The Architecture of Imperialism: Military Bases and the Evolution of Foreign Policy in Egypt's New Kingdom* (Leiden and Boston: Brill; 2005).

Murnane, William. "Once Again the Dates for Thutmose III and Amenhotep III," *JANES* 3 (1970/71): 1–7.

———. "On the Accession Date of Akhenaten," in Janet Johnson and Edward Wente (eds.), *Studies in Honor of George R. Hughes* (Chicago: Oriental Institute; 1976), 163–67.

———. "The Sed Festival: A Problem in Historical Method," *MDAIK* 37 (1981): 369–76.

———. *The Road to Kadesh: A Historical Interpretation of the Battle Reliefs of King Sety I at Karnak* (Chicago: Oriental Institute; 1985).

———. "The Kingship of the Nineteenth Dynasty: A Study in the Resilience of an Institution," in David O'Connor and David Silverman (eds.), *Ancient Egyptian Kingship* (Leiden, New York, and Cologne: Brill: 1995), 199–200.

Murnane, William, and Van Siclen III, Charles. *The Boundary Stelae of Akhenaton* (London and New York: Kegan Paul International; 1993).

Murray, Margaret A. *The Osireion at Abydos* (London: B. Quaritch; 1904).

Naville, Edouard. *The XIth Dynasty Temple of Deir el-Bahari* I (London: Egypt Exploration Fund; 1907).

Neugebauer, Otto. "Die Bedeutungslosigkeit der 'Sothisperiode' für die älteste ägyptische Chronologie," *Acta Orientalia* 17 (1939): 169–85.

———. "The Origin of the Egyptian Calendar," *JNES* 1 (1942): 396–403.

———. The 'Metonic Cycle' in Babylonian Astronomy. Studies in Ancient Astronomy. VI.," in M. F. Ashley-Montagu (ed.), *Studies and Essays in the History of Science and Learning Offered in Homage to George Sarton. On the Occasion of his Sixtieth Birthday, 31 August 1944* (New York: Henry Schuman; 1946), 435–47.

———. "The Water Clock in Babylonian Astronomy," *Isis* 37 (1947): 37–43.

———. *The Exact Sciences in Antiquity*[2] (New York: Dover; 1969).

———. *Astronomical Cuneiform Texts* III (New York, Heidelberg, and Berlin: Springer Verlag; 1983).

Neugebauer, Otto, and Parker, Richard A. *Egyptian Astronomical Texts I: The Early Decans* (Providence and London: L. Humphries; 1960).

———. *Egyptian Astronomical Texts II: The Ramesside Star Clocks* (Providence and London: L. Humphries; 1964).

———. *Egyptian Astronomical Texts III: Decans, Planets, Constellations, and Zodiacs* (London: Lund Humphries; 1969).

Nilsson, Martin. *Primitive Time Reckoning* (Lund: C. W. K. Gleerup; 1920).

Nims, Charles F. "An Oracle in the 'Repeating of Births,'" *JNES* 7 (1948): 157–62.

———. "Ramesseum Sources of Medinet Habu Reliefs," in Janet H. Johnson and Edward F. Wente (eds.), *Studies in Honor of George R. Hughes: January 12, 1977* (Chicago: Oriental Institute; 1976), 169–75.

Nolan, John S. "Lunar Intercalations and 'Cattle Counts' during the Old Kingdom: The Hebsed in Context," in Hana Vymazalová and Miroslav Bárta (eds.), *Chronology and Archaeology in Ancient Egypt: The Third Millennium B.C.* (Prague: Czech Institute of Egyptology; 2009), 44–60.

North, Robert. "Phoenicia-Canaan Frontier Lᵉbô' of Ḥama," *Mélanges de l'Université St. Joseph* 46 (1970–71): 71–103.

Obsomer, Claude. "La bataille de Qadech de Ramsès II: les *n'arin, sekou tepy* et questions d'itinéraires," in Christina Karlshausen and Claude Obsomer (eds.), *De la Nubie à Qadech / From Nubia to Qadesh. Études sur la guerre dans l'Égypte ancienne* (Brussels: Safran; 2016), 81–170.

O'Connor, David. "The Sea Peoples and the Egyptian Sources," in Eliezer D. Oren (ed.), *The Sea Peoples and their World: A Reassessment* (Philadelphia: University Museum; 2000), 85–102.

Odorico, Marco de. *The Use of Numbers and Quantifications in the Assyrian Royal Inscriptions* (Helsinki: Neo-Assyrian Text Corpus Project; 1995).

O'Mara, Patrick. "Censorinus, the Sothic Cycle, and Calendar Year One in Ancient Egypt: The Epistemological Problem," *JNES* 62 (2003): 17–26.

Oosthoek, Ann-Laurie. "Hittite ou pas Hittite? Trois représentations à caractère hybride," in Claude Obsomer and Ann-Laure Oosthoek (eds.), *Amosiadès: Mélanges offerts au professeur Claude Vandersleyen par ses anciens étudiants* (Louvain-la-Neuve: Université catholique de Louvain, Institut orientaliste, Collège Erasme; 1992), 335–46.

Oren, Eliezer D. "Ancient Military Roads between Egypt and Canaan," *Bulletin of the Anglo-Israel Archaeological Society* 10 (1982/1983): 20–24.

———. "Migdol: A New Fortress on the Edge of the Eastern Nile Delta," *BASOR* 256 (1984): 7–44.

———. "The 'Ways of Horus' in North Sinai," in Anson F. Rainey (ed.), *Egypt, Israel, Sinai. Archaeological and Historical Relationships in the Biblical Period* (Tel Aviv: Tel Aviv University; 1987), 69–119.

———. "The Establishment of Egyptian Imperial Administration on the 'Ways of Horus:' An Archaeological Perspective from North Sinai," in Ernst Czerny et al. (eds.), *Timelines: Studies in Honour of Manfred Bietak* II (Leuven, Paris, and Dudley MA: Peeters; 2006), 279–92.

Oren, Eliezer D., and Shereshevski, J. "Military Architecture along the 'Ways of Horus' – Egyptian Reliefs and Archaeological Evidence," *Eretz Israel* 20 (1989): 8–22.

Parker, Richard A. "Darius and His Egyptian Campaign," *AJSL* 58 (1941): 373–77.

———. "Persian and Egyptian Chronology," *AJSL* 58 (1941): 285–301.

———. "Ancient Jewish Calendation: A Criticism," *JBL* 63 (1944): 173–76.

———. *The Calendars of Ancient Egypt* (Chicago: University of Chicago; 1950).

———. Review of Siegfried Schott, *Altägyptische Festdaten.* Akademie der Wissenschaften und der Literatur, Abhandlungen der Geistes- und Sozialwissenschaftlichen Klasse 1950 (Wiesbaden: F. Steiner; 1950.10), *BiOr* 9 (1952): 100–102.

———. "Sothic Dates and Calendar 'Adjustment,'" *RdE* 9 (1952): 102–8.

———. "The Year in Ancient Egypt," *Journal of Calendar Reform* 20 (1955): 81–83.

———. "The Calendar: Past, Present, and Future," *Round Table* (University of Chicago: An NBC Radio Discussion) 873 (January 1955): 1–12.

———. "The Lunar Dates of Thutmose III and Ramesses II," *JNES* 16 (1957): 39–43.

———. "The Problem of the Month-Names: A Reply," *RdE* 11 (1957): 85–107.

———. "Once Again the Coregency of Thutmose III and Amenhotep II," in E. B. Hauser (ed.), *Studies in Honor of John A. Wilson* (Chicago: 1969), 75–82.

———. "The Beginning of the Lunar Month in Ancient Egypt," *JNES* 29 (1970): 217–20.

———. "The Calendars and Chronology," in J. R. Harris (ed.), *The Legacy of Egypt*[2] (Oxford: Clarendon; 1971), 13–26.

———. "The Sothic Dating of the Twelfth and Eighteenth Dynasties," in Janet H. Johnson and Edward F. Wente (eds.), *Studies in Honor of George R. Hughes: January 12, 1977* (Chicago: University of Chicago; 1976), 177–89.

———. "Some Reflections on the Lunar Dates of Thutmose III and Ramesses II," in William K. Simpson and Whitney M. Davis (eds.), *Studies in Ancient Egypt, the Aegean, and the Sudan: Essays in Honor of Dows Dunham on the Occasion of his 90th Birthday, June 1, 1980* (Boston: Museum of Fine Arts; 1981), 146–48.

Parker, Richard A., and Dubberstein, Waldo H. *Babylonian Chronology 626 B.C.–A.D. 45* (Chicago: University of Chicago Press; 1942).

Pasquali, Stéphane. "La date du Papyrus BM 10056: Thoutmosis III ou Amenhotep II?," *RdE* 58 (2007): 71–86.

Porceddu, Sebastian; Jetsu, Lauri; Merkkanen, Tapio; and Toivari-Viitala, Jaana, "Evidence of Periodicity of Ancient Egyptian Calendars of Lucky and Unlucky Days," *CAJ* 18 (2008): 327–39.

Posener, Georges. *Catalogue des ostraca hiératiques de Deir el Médineh* I (Cairo: Institut Français d'Archéologie Orientale; 1938).

———. *Cinq figurines d'envoûtement* (Cairo: Institut Français d'Archéologie Orientale; 1987).

Posener-Kriéger, Paule. Review of Ann Macy Roth, *Egyptian Phyles in the Old Kingdom: The Evolution of a System of Social Organization* (Chicago: Oriental Institute; 1991), *CdE* 71 (1996): 73–79.

Pritchett, William. "The Athenian Lunar Month," *Classical Philology* 54 (1959): 151–57.

———. "Calendars of Athens Again," *BCH* 81 (1957): 269–301.

———. "The Calendar of the Athenian Civic Administration," *Phoenix* 30 (1976): 337–56.

———. "The Calendar of the Gibbous Moon," *ZPE* 49 (1982) 243–66.

———. "The Solar Year of Thucydides," in William Pritchett, *Thucydides' Pentekontaetia and Other Essays* (Amsterdam: J. C. Gieben; 1995), 173–204.

———. "Postscript: The Athenian Calendar," *ZPE* 128 (1999): 79–93.

Pritchett, William, and Neugebauer, Otto. *The Calendars of Athens* (Cambridge, MA: Harvard University Press; 1947).

Puhvel, Jaan. "Eventide in Hatti-land," in Brigitte L. M. Bauer and Georges-Jean Pinault (eds.), *Language in Time and Space. A Festschrift for Werner Winter on the Occasion of his 80th Birthday* (Berlin and New York: Mouton de Gruyter; 2003), 347–49.

Quack, Joachim. *Untersuchungen zu den ägyptischen Dekanen und ihrer Rezeption in der griechisch-römischen Welt* (Habilitationsschrift, Berlin: Freie Universität; 2002 submitted, accepted 2003).

Quirke, Stephen. "Frontier or Border? The Northeast Delta in Middle Kingdom Texts," in Alexandra Nibbi (ed.), *The Archaeology, Geography and History of the Egyptian Delta in Pharaonic Times.* Discussions in Egyptology Special Number 1 (Oxford: Cotswold Press; 1989), 261–79.

Raulwing Peter, and Clutton-Brock, Juliet. "The Buhen Horse: Fifty Years after its Discovery (1958–2008)," *JEGH* 2 (2009): 1–106.

Redford, Donald. *Pharaonic King-Lists, Annals and Day Books: A Contribution to the Study of the Egyptian Sense of History* (Mississauga: Benben Publications; 1986).

———. *The Wars in Syria and Palestine of Thutmosis III* (Leiden and Boston: Brill; 2003).

Reisner, George A. "The Tomb of Hepzefa, Nomarch of Siût," *JEA* 5 (1918): 79–98.

Riaud, Jean (ed.). *Étrangers et exclus dans le monde biblique. Colloque international à l'Université Catholique de l'Ouest. Angers, les 20 et 21 février 2002.* Théolarge 3 (Angers: L.T.S.R. Laboratoire de théologie et des sciences religieuses; 2003).

Ricke, Herbert; Hughes, George R.; and Wente, Edward. *The Beit el-Wali Temple of Ramesses II* (Chicago: Oriental Institute; 1967).

Rittner, Robert. "Libyan vs. Nubian as the Ideal Egyptian," in Stephen E. Thompson and Peter der Manuelian (eds.), *Egypt and Beyond: Essays Presented to Leonard H. Lesko upon his Retirement from the Wilbour Chair of Egyptology at Brown University, June 2005* (Providence: Dept. of Egyptology and Ancient Western Asian Studies; 2008), 305–14.

Roccati, Alessandro. "La conception rituelle du désert chez les anciens égyptiens," in Y. Christe et al. (eds.), *Le désert: Image et réalité. Actes du Colloque de Cartigny 1983* (Leuven: Peeters; 1983), 127–29.

Rose, Lynn E. *Sun, Moon, and Sothis: A Study of Calendar Reforms in Ancient Egypt* (Deerfield Beach, FL: Kronos Press; 1999).

Roth, Ann Macy. *Egyptian Phyles in the Old Kingdom: The Evolution of a System of Social Organization* (Chicago: Oriental Institute; 1991).

Rössler, Otto. *Gesammelte Schriften zur Semitohamitistik*, eds. Thomas Schneider and Oskar Kaelin (Münster: Ugarit Verlag; 2001).

Rüpke, Jörg. *Kalender und Öffentlichkeit. Die Geschichte der Repräsentation und religiösen Qualifikation von Zeit in Rom* (Berlin and New York: Walter de Gruyter; 1995).

Saad, Ramadan, and Manniche, Lise. "A Unique Offering List of Amenophis IV Recently Found at Karnak," *JEA* 57 (1971): 70–72.

Sachs, Abraham. "Sirius Dates in Babylonian Astronomical Texts of the Seleucid Period," *JCS* 6 (1952): 105–14.

Sanchez, Gonzalo M. "Variations of Representation in the Direction of the Battle of Kadesh," *JSSEA* 31 (2004): 129–50.

Sauneron, Serge. *Les fêtes religieuses d'Esna* (Cairo: Institut Français d'Archéologie Orientale; 1962).

Schaefer, Bradley. "Visibility of the First Lunar Crescent," *Quarterly Journal of the Royal Astronomical Society* 29 (1988): 511–23.

———. "Astronomy and the Limits of Vision," *Vistas in Astronomy* 36 (1993): 311–61.

———. "Lunar Crescent Visibility," *Quarterly Journal of the Royal Astronomical Society* 37 (1996): 759–68.

———. "Astronomy in Historical Studies," *Archaeoastronomy* 14 (1999): 89–108.

———. "New Methods and Techniques for Historical Astronomy and Archaeoastronomy," *Archaeastronomy* 15 (2000): 121–36.

———. "The Heliacal Rise of Sirius and Ancient Egyptian Chronology," *Journal for the History of Astronomy* 31 (2000): 149–55.

Schaefer, Bradley and Hillier, William. "Refraction Near the Horizon," *Publications of the Astronomical Society of the Pacific* 102 (2000): 796–805.

Schäfer, Heinrich. "Ein Bruchstück altägyptischer Annalen," *Abhandlungen der Königlichen Preussischen Akademie der Wissenschaften zu Berlin: Philos.-hist. Klasse, Anhang Abhandlungen* 1 (1902).

Schapiro, Meyer. *Words and Pictures. On the Literal and the Symbolic in the Illustration of a Text* (The Hague and Paris: Mouton; 1973).

Schlott-Schwab, Adelheid. *Die Ausmasse Ägyptens nach altägyptischen Texten* (Wiesbaden: Harassowitz; 1981).

Schulman, Alan Richard. *Military Rank, Title and Organization in the Egyptian New Kingdom* (Berlin: Bruno Hessling; 1964).

Scolnic, Benjamin Edidin. "A New Working Hypothesis for the Identification of Migdol," in James K. Hoffmeier and Alan Millard (eds.), *The Future of Biblical Archaeology: Reassessing Methodologies and Assumptions* (Grand Rapids: Eerdmans; 2004), 91–120.

Schott, Siegfried. *Altägyptische Festdaten.* Akademie der Wissenschaften und der Literatur, Abhandlungen der Geistes- und Sozialwissenschaftlichen Klasse 1950.10 (Wiesbaden: F. Steiner; 1950).

———. *Das schöne Fest vom Wüstentale: Festbräuche einer Totenstadt.* Akademie der Wissenschaften und der Literatur, Abhandlungen der Geistes- und Sozialwissenschaftlichen Klasse 1952.11 (Wiesbaden: F. Steiner; 1953).

Scott, Nora. "Recent Additions to the Egyptian Collection," *BMMA* 15 (NS) (1956): 79–92.

Servajean, Frédéric. *Quatre études sur la bataille de Qadech.* CENiM 6 (Montpellier: Université Paul Valéry (Montpellier III), Équipe "Égypte Nilotique et Méditerranéenne"; 2012).

Sethe, Kurt. *Von Zahlen und Zahlworten bei den alten Ägyptern und was für andere Völker und Sprachen daraus zulernen ist: Ein Beitrag zur Geschichte von Rechenkunst und Sprache* (Strasburg: Trübner; 1916).

Shafer, Byron E. "Temples, Priests, and Rituals: An Overview," in Byron E. Schafer (ed.), *Temples of Ancient Egypt* (Ithaca; Cornell University Press; 1997), 1–30.

Shai, Itzick, et al. "The Importance of the Donkey as a Pack Animal in the Early Bronze Age Southern Levant: A View from Tell es Safi/Gath," *ZDPV* 132 (2016): 1–25.

Sloley, Robert Walter. "The Origin of the 365-Day Egyptian Calendar," *ASAE* 48 (1948): 261–65.

Spalinger, Anthony. "Notes on the Reliefs of the Battle of Kadesh," in Hans Goedicke (ed.), *Perspectives on the Battle of Kadesh* (Baltimore: Halgo; 1985), 1–42.

———. "The Grain System of Dynasty 18," *SAK* 14 (1987): 283–311.

———. "A Remark on Renewal," SAK 17 (1990): 289–94.

———. Review of Christian Leitz, *Studien zur ägyptischen Astronomie* (Wiesbaden: Harrassowitz; 1989), *OLZ* 87 (1992): 23–26.

———. "Some Remarks on the Epagomenal Days in Ancient Egypt," *JNES* 54 (1994): 33–47.

———. "Calendars: Real and Ideal," in Betsy Bryan and David Lorton (eds.), *Essays in Egyptology in Honor of Hans Goedicke* (San Antonio: Van Siclen Press; 1994), 297–308.

———. (ed.). *Revolutions in Time: Studies in Ancient Egyptian Calendrics* (San Antonio: Van Siclen Press; 1994).

———. "Under the Moon of Earth," in Anthony Spalinger (ed.), *Revolutions in Time: Studies in Ancient Egyptian Calendrics* (San Antonio: Van Siclen Books; 1994), 61–83.

———. "Thoth and the Calendars," in Anthony Spalinger (ed.), *Revolutions in Time: Studies in Ancient Egyptian Calendrics* (San Antonio: Van Siclen Books; 1994), 45-60.

———. "Calendrical Comments," *BiOr* 51 (1994): 5–20.

———. "Parallelism of Thought," in C. Berger, G. Clerc, and N. Grimal (eds.), *Hommages à Jean Leclant* IV (Cairo: Institut Français d'Archéologie Orientale; 1994), 363–77.

———. "Notes on Ancient Egyptian Calendars," *Or* 64 (1995): 17–32.

———. "The Calendrical Importance of the Tombos Stela," *SAK* 22 (1995): 271–81.

———. "Month Representations," *CdÉ* 70 (1995): 110–22.

———. "The Lunar System of Festival Calendars: From the New Kingdom Onwards," *BSEG* 19 (1995): 25–40.

———. *The Private Feast Lists of Ancient Egypt* (Wiesbaden: Harrassowitz: 1996).

———. "Some Times," *RdE* 47 (1996): 67–77.

———. "The Festival Structure of Thutmose III's Buto Stela," *JARCE* 33 (1996): 69–76.

———. "Sothis and 'Official' Calendar Texts," *Varia Aegyptia* 10 (1995 = 1997), 175–83.

———. “Chronological Remarks,” *BSEG* 22 (1998): 51–58.

———. “Praise God and Pay the Priests,” *ÄuL* 7 (1998): 43–57.

———. Review of Lynn E. Rose, *Sun, Moon, and Sothis: A Study of Calendar Reforms in Ancient Egypt* (Deerfield Beach, FL: Kronos Press; 1999), *Bryn Mawr Classical Review* 2000.01.17, http://bmcr.brynmawr.edu/2000/2000-01-17.html.

———. “Egyptian Festival Dating and the Moon,” in John M. Steele and Annette Imhausen (eds.), *Under One Sky: Astronomy and Mathematics in the Ancient Near East* (Münster: Ugarit Verlag; 2002), 379–403.

———. *The Transformation of an Ancient Egyptian Narrative: P. Sallier III and the Battle of Kadesh* (Wiesbaden; Harrassowitz; 2002).

———. “The Battle of Kadesh: The Chariot Frieze at Abydos,” *ÄuL* 13 (2003): 163–99.

———. *War in Ancient Egypt* (Malden, Oxford, and Carlton: Blackwell; 2005).

———. “Land Brightens, Day Begins,” in Anthony Spalinger, *Five Views on Egypt* (Göttingen: Seminar für Ägyptologie und Koptologie; 2006).

———. “Some Notes on the Chariot Arm of Egypt in the Early New Kingdom,” in Panagiotis Kousoulis and Konstantinos Magliveras (eds.), *Moving Across Borders: Foreign Relations, Religion, and Cultural Interactions in the Ancient Mediterranean* (Leuven, Paris, Dudley: Peeters, 2007), 119–37.

———. “Ramesses Ludens et Alii,” in Panagiotis Kousoulis (ed.), *Studies on the Ancient Egyptian Culture & Foreign Relations* (Rhodes: University of the Aegean; 2007), 71–86.

———. “Re-Reading Egyptian Military Reliefs,” in Mark Collier and Steven Snape (eds.), *Ramesside Studies in Honour of K.A. Kitchen* (Bolton: Rutherford Press; 2011), 475–91.

———. Review of John M. Steele (ed.), *Calendars and Years: Astronomy and Time in the Ancient Near East* (Oxford: Oxbow; 2007), *JAOS* 131 (2011): 311–13.

———. “The Beginning of the Civil Calendar,” in Miroslav Bárta, Filip Coppens, and Jaromír Krejčí (eds.), *Abusir and Saqqara in the Year 2010* (Prague: Czech Institute of Egyptology; 2012).

———. “The Date of Amunhotep II’s First Accession,” *SAK* 40 (2011): 387–97.

———. *Icons of Power: A Strategy of Reinterpretation* (Prague: Faculty of Arts, Charles University in Prague; 2012).

———. “Nut and the Egyptologists,” *SAK* 41 (2012): 53–77.

———. “Divisions in Monumental Texts and Their Images: The Issue of Kadesh and Megiddo,” in Mayer I. Gruber et al. (eds.), *All the Wisdom of the East: Studies in Near Eastern Archaeology and History in Honor of Eliezer D. Oren* (Fribourg and Göttingen: Academic Press and Vandenhoeck & Ruprecht; 2012), 373–93.

———. "Egyptian and Greek Time Frames: The Date of the Kronia Festival," in Miroslav Bárta and Hella Küllmer (eds.), *Diachronic Trends in Ancient Egyptian History. Studies Dedicated to the Memory of Eva Pardey* (Prague: Philosophy Faculty, Charles University; 2013), 109–18.

———. "Eleventh Day, Twelfth Night: Further Remarks Concerning Three Feasts in Egyptian Civil Thoth," *SAK* 43 (2014): 399–415.

———. Review of Jan Assmann, *Steinzeit und Sternzeit: Altägyptische Zeitkonzepte* (Munich: Wilhelm Fink; 2011), *WZKM* 102 (2013): 303–9.

———. "Operational Bases: Gaza and Beth Shan," in Christina Karlshausen and Claude Obsomer (eds.), De la Nubie à Qadech / From Nubia to Qadesh. Études sur la guerre dans l'Égypte ancienne (Brussels: Safran; 2016), 63–80.

———. "The Cycle of 309 Egyptian Months in the Amduat," in Camilla Di Biase-Dyson and Leonie Donovan (eds), *The Cultural Manifestations of Religious Experience. Studies in Honour of Boyo G. Ockinga* (Münster: Ugarit Verlag; 2017), 377–82.

Speelers, Louis. *Le papyrus de Nefer Renpet: un livre des morts de la XVIIIe Dynastie aux Musées royaux du cinquantenaire à Bruxelles* (Brussels: Vromant & Co.; 1917).

Spruytte, Jérôme. *Early Harness Systems: Experimental Studies. A Contribution to the History of the Horse,* trs. Mary Littauer (London: J. Allen; 1983).

———. "Le char antique du Musée Archéologique de Florence," in *Travaux de LAMPO* 1989, 185–90.

———. *Attelages antiques libyens: archéologie saharienne expérimentale* (Paris: Maison des sciences de l'homme; 1996).

———. "L'attelage égyptien sous la XVIIIe dynastie: étude technique et technologique," *Kyphi* 2 (1999): 77–87.

Stager, Lawrence E. "Chariot Fittings from Philistine Ashkelon," in Seymour Gittin, J. Edward Wright, and J. P. Dessel (eds.), *Confronting the Past: Archaeological and Historical Essays on Ancient Israel in Honor of William G. Dever* (Winona Lake: Eisenbrauns; 2006), 170–76.

Staubli, Thomas. *Das Image der Nomaden im alten Israel und in der Ikonographie seiner sesshaften Nachbarn* (Göttingen and Fribourg: Universitätsverlag and Vandenhoeck & Ruprecht; 1991).

Steele, John M. "The Length of the Month in Mesopotamian Calendars of the First Millennium BC," in John M. Steele (ed.), *Calendars and Years: Astronomy and Time in the Ancient Near East* (Oxford: Oxbow; 2007), 133–48.

Stefanovic, Danijela. "The Middle Kingdom Chronology: Present Knowledge and Confusions," *Journal of the Serbian Archaeological Society* 17 (2001): 91–102.

Stewart, H. M. *Egyptian Stelae, Reliefs and Paintings from the Petrie Collection* (Warminster: Aris and Phillips; 1979).

Tefnin, Roland. "Image, écriture, récit. À propos des représentations de la bataille de Qadesh," *GM* 47 (1981): 55–78.

———. "Discours et iconicité dans l'art égyptien," *GM* 79 (1984): 55–71.

———. "La perception de la différence en Égypte pharaonique," *Civilisations* 35 (1986): 39–55.

———. "Élements pour une sémiologie de l'image égyptienne," *CdE* 66 (1991): 60–88.

Trapani, Marcella. "Une nouvelle enquête sur la stèle d'Ahmès-Néfertari," *ZÄS* 129 (2002): 152–65.

Traunecker, Claude. "Données nouvelles sur le début du règne d'Aménophis IV et son œuvre à Karnak," *JSSEA* 14 (1984): 60–69.

———. "Amenhotep IV, percepteur royal du Disque," in Thierry-Louis Bergerot (ed.), *Akhénaton et l'époque amarnienne* (Paris: Edition Chéops and Centre d'Égyptologie; 2005), 145–82.

Turner, Eric G., and Neugebauer, Otto. "Gymnasium Debts and New Moons," *BJRL* 32 (1949): 80–96.

Tylor, Joseph John. *Wall Drawings and Monuments of El Kab: The Tomb of Paheri* (London: Egypt Exploration Fund; 1895).

Valbelle, Dominique. "Entre l'Égypte et la Palestine, Tell el-Herr," *BSFE* 109 (1987): 24–38.

Vandersleyen, Claude. Review of G. A. Gaballa, *Narrative in Egyptian Art* (Mainz am Rhein: Philipp von Zabern; 1976), *CdE* 53 (1978): 275–78.

Vassy, E. "La lumière du ciel nocturne," *BIE* 33 (1950–51): 203–6.

Vaumas, Étienne de. "Sur le mouvement tournant effectué par Ramsès II à la veille de la bataille de Qadesh," *Mélanges de l'Université St. Joseph* 46 (1970–71): 53–67.

Verner, Miroslav. "Contemporaneous Evidence for the Relative Chronology of Dyns. 4 and 5," in Erik Hornung, Rolf Krauss, and David A. Warburton (eds.), *Ancient Egyptian Chronology* (Leiden and Boston: Brill; 2006), 124–43.

———. "The System of Dating in the Old Kingdom," in Hana Vymazalová and Miroslav Bárta (eds.), *Chronology and Archaeology in Ancient Egypt: The Third Millennium B.C.* (Prague: Czech Institute of Egyptology; 2009), 23–43.

Vernus, Pascal. "Un texte oraculaire de Ramsès VI," *BIFAO* 73 (1975): 103–10.

———. "Les espaces de l'écrit dans l'Égypte pharaonique," *BSFE* 119 (1990): 39–43.

———. "Réception linguistique et idéologique d'une nouvelle technologie: le cheval dans la civilisation pharaonique," in Myriam Wissa (ed.), *The Knowledge Economy and Technological Capabilities: Egypt, the Near East and the Mediterranean, 2nd millennium B.C.–1st millennium A.D.* (Barcelona: Editorial AUSA; 2009), 1–46.

Vita, Juan-Pablo. "The Power of a Pair of War Chariots in the Late Bronze Age: On Letters RS 20.33 (Ugarit), BE 17.33a (Nippur), and EA 197 (Damascus Region)," in Jordi Vidal (ed.), *Studies on War in the Ancient Near East: Collected Essays on Military History* (Münster: Ugarit Verlag; 2010), 87–93.

Vogel, Carola. "Die Schlacht von Megiddo: Thutmose III. im Kampf um Syrien und Palästina," *Ished* (Journal des Ägypten Forum Berlin e.V.) 1/2001, 15–28.

Voß, Susanne. *Untersuchungen zu den Sonnenheiligtümern der 5. Dynastie: Bedeutung und Funktion eines singulären Tempeltyps im Alten Reich* (Hamburg: PhD Dissertation: University of Hamburg; 2004).

Vymazlová, Hana. "Some Remarks on the *wꜣg*-Festival in the Papyrus Archive of Raneferef," in Hana Vymazlová and Miroslav Bárta (eds.), *Chronology and Archaeology in Ancient Egypt (The Third Millennium B.C.)* (Prague: Czech Institute of Egyptology; 2008), 137–43.

Walsem, R. van. "Month-Names and Feasts at Deir el-Medîna," in R. J. Demarée and Jac. J. Janssen (eds.), *Gleanings from Deir el Medîna* (Leiden: Nederlands Instituut voor het Nabije Oosten; 1982), 215–44.

Ward, William. "The Shasu 'Bedouin:' Notes on a Recent Publication," *JESHO* 15 (1972): 35–60.

Way, Thomas von der. *Die Textüberlieferung Ramses' II. zur Qadeš-Schlacht* (Hildesheim: Gerstenberg Verlag; 1984).

———. *Göttergericht und 'Heiliger' Krieg im Alten Ägypten: Die Inschriften des Merenptah zum Libyerkrieg des Jahres 5* (Heidelberg: Heidelberger Orientverlag; 1992).

Webster, Hutton. *Rest Days: A Study in Early Law and Morality* (New York: Macmillan; 1916).

Webb, Ruth. "*Ekphrasis* Ancient and Modern: The Invention of a Genre," *Word and Image* 15 (1999): 7–18.

Wells, Ronald. "Some Astronomical Reflections on Parker's Contributions to Egyptian Chronology," in Leonard H. Lesko (ed.), *Egyptological Studies in Honor of Richard A. Parker* (Hanover and London: University Press of New England; 1986); 165–71.

——. Review of Christian Leitz, *Studien zur ägyptischen Astronomie* (Wiesbaden: Harrassowitz; 1989), *BiOr* 49 (1992): 721–28.

——. Review of Lynn E. Rose, *Sun, Moon, and Sothis: A Study of Calendar Reforms in Ancient Egypt* (Deerfield Beach, FL: Kronos Press; 1999), *JNES* 61 (2002): 311–15.

——. "The Role of Astronomical Techniques in Ancient Egyptian Chronology: The Use of Lunar Month Lengths in Absolute Dating," in John M. Steele and Annette Imhausen (eds.), *Under One Sky: Astronomy and Mathematics in the Ancient Near East* (Münster; Ugarit Verlag; 2002), 459–72.

Wente, Edward. "Thutmose III's Accession and the Beginning of the New Kingdom," *JNES* 34 (1975): 265–72.

Whitehead, Alfred North, and Russell, Bertrand. *Principia Mathematica* (Cambridge: Cambridge University Press; 1910–13).

Wiebach, Silvia. "Die Begegnung von Lebenden und Toten im Rahmen des thebanischen Talfestes," *SAK* 13 (1986): 263–91.

Wikgren, Heidi. "The Festival Calendar at Deir el-Medina," in Kathryn Piquette and Serena Love (eds.), *Current Research in Egyptology 2003* (Oxford: Oxbow; 2005), 179–200.

Wilkinson, Toby A. H. *Royal Annals of Ancient Egypt: The Palermo Stone and its Associated Fragments* (London and New York: Kegan Paul International; 2000).

Willems, Harco. *The Coffin of Heqata (Cairo JdE 36418): A Case Study of Egyptian Funerary Culture of the Early Middle Kingdom* (Leuven: Peeters; 1996).

Winand, Jean. "Un mystère enfin éclairci!," in *"Wo ist denn hier der Satz?" Fastschrift für Friedrich Junge überreicht zu seinem 60. Geburtstag am 18. April 2001 (= Lingua Aegyptia* 9 ½ [2001]).

Winlock, Herbert. *The Rise and Fall of the Middle Kingdom at Thebes* (New York: Macmillan; 1947).

Wit, Constant de. "Inscriptions dédicatoires du temple d'Edfou," *CdE* 36 (1961), 56–97 and 277–320.

Wittgenstein, Ludwig. *Tractatus Logico-Philosophicus,* with an introduction by Bertrand Russell (London: Routledge and Keagen; 1922).

Wohlgemuth, Gottfried. *Das Sokarfest* (Göttingen, PhD Dissertation: University of Göttingen; 1957).

Wolf, Walther. Review of William Stevenson Smith, *The Art and Architecture of Ancient Egypt* (Harmondsworth: Penguin; 1959), *OLZ* 56 (1961): 17–19.

Yallop, B. D. *A Method for Predicting the First Sighting of the New Crescent Moon.* (HM Nautical Almanac Office NAO Technical Note 69) (Chilton: Council for the Central Laboratory of the Research Councils; June 1997 and updated April 1998).

Yoyotte, Jean. "La date supposée du couronnement d'Hatchepsout," *Kêmi* 18 (1968): 85–91.

Yoyotte, Jean, and Lopez, Jésus. "L'organisation de l'armée et les titulaires de soldats au nouvel empire égyptien," *BiOr* 26 (1969): 3–19.

Yurko, Frank. "Merenptah's Canaanite Campaign," *JARCE* 23 (1986): 189–215.

———. "Merenptah's Canaanite Campaign and Israel's Origins," in Ernest S. Frerichs and Leonard H. Lesko (eds.), *Exodus: The Egyptian Evidence* (Winona Lake: Eisenbrauns; 1997), 27–55.

Zanker, Graham. *Modes of Viewing in Hellenistic Poetry and Art* (Madison: University of Wisconsin Press; 2004).

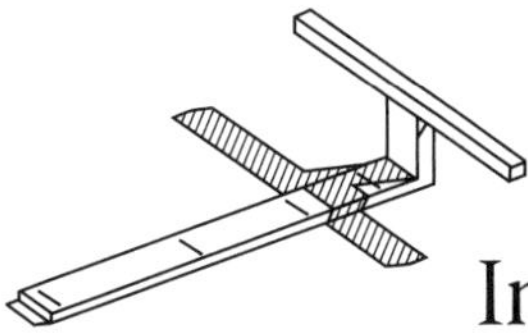

# Index

*Note that the terms "calendar" and "Civil Calendar" are not included.*

EGYPTIAN WORDS



LATIN WORDS